大数据与人工智能技术丛书

大数据技术入门
——Hadoop+Spark

◎ 于海浩 刘志坤 主编　韩咏 孙栩 副主编

清华大学出版社

北京

内 容 简 介

本书基础理论、应用开发以及实际案例相结合，围绕 Hadoop、Spark 生态圈循序渐进地介绍关于大数据技术领域中的基础知识、应用开发技术和基于 Spark 的常见机器学习算法，最后以两个实战案例全面、系统地应用了本书介绍的基础知识和应用开发方法。全书共 14 章，分别为大数据概述、Hadoop 简介及安装部署、HDFS、MapReduce 计算框架、Hive 数据仓库、HBase 分布式数据库、Spark 基础、Spark RDD 弹性分布式数据集、Spark SQL、Spark Streaming 实时计算框架、Spark Streaming 与 Flume、Kafka 的整合、Spark MLlib 机器学习、实战案例——分布式优惠券后台应用系统和实战案例——新闻话题实时统计分析系统，书中的每个知识点都有相应的实现代码和实例。

本书主要面向广大从事大数据分析、应用开发、机器学习、数据挖掘的专业人员以及从事高校信息技术专业的教师和高等院校的在读学生及相关领域的广大科研人员。

本书封面贴有清华大学出版社防伪标签，无标签者不得销售。
版权所有，侵权必究。举报：010-62782989，beiqinquan@tup.tsinghua.edu.cn。

图书在版编目（CIP）数据

大数据技术入门：Hadoop＋Spark/于海浩，刘志坤主编．—北京：清华大学出版社，2022.1(2023.3重印)
（大数据与人工智能技术丛书）
ISBN 978-7-302-59181-8

Ⅰ．①大… Ⅱ．①于… ②刘… Ⅲ．①数据处理软件 Ⅳ．①TP274

中国版本图书馆 CIP 数据核字(2021)第 187094 号

责任编辑：陈景辉　张爱华
封面设计：刘　键
责任校对：胡伟民
责任印制：丛怀宇

出版发行：清华大学出版社
网　　址：http://www.tup.com.cn, http://www.wqbook.com
地　　址：北京清华大学学研大厦 A 座　　邮　编：100084
社 总 机：010-83470000　　邮　购：010-62786544
投稿与读者服务：010-62776969, c-service@tup.tsinghua.edu.cn
质量反馈：010-62772015, zhiliang@tup.tsinghua.edu.cn
课件下载：http://www.tup.com.cn, 010-83470236

印 装 者：三河市天利华印刷装订有限公司
经　　销：全国新华书店
开　　本：185mm×260mm　　印　张：13　　字　数：325 千字
版　　次：2022 年 1 月第 1 版　　印　次：2023 年 3 月第 4 次印刷
印　　数：4001～5500
定　　价：49.90 元

产品编号：091468-01

前　言

近年来，随着大数据底层设施的逐渐成熟，大数据技术开始结合具体行业，向行业应用延伸。围绕数据资源、基础硬件、通用软件等方面的大数据产业正在逐渐形成。各行各业对大数据人才的需求也日益增强。鉴于此，国内外一些高校先后开设"数据科学与大数据"专业，旨在培养具备大数据技术的高级人才。

为满足相关技术人员和高校师生学习大数据技术基础知识，我们在总结近几年大数据技术基础知识和应用案例的基础上，以理论结合实践的方式将大数据基本概念、大数据开发技术与实际应用相结合，精心组织并完成了本书的编写。

本书主要内容

本书为一本夯实大数据基础知识，以实际应用为导向的书籍，非常适合初、中级学习大数据技术的读者。读者可以在短时间内学习本书中介绍的所有知识，掌握大数据技术的开发方法。

作为一本关于大数据技术的入门书籍，本书共有 14 章。

第 1 章主要介绍了大数据的研究背景、大数据的定义及其技术特点以及大数据处理的主要技术特点与难点，最后阐述了研究大数据的意义。

第 2 章主要介绍了 Hadoop 的起源、生态体系和集群架构，对 Hadoop 的安装配置进行了详细的讲解。

第 3 章主要介绍了 Hadoop 中非常重要的分布式存储文件系统——HDFS，分析了 HDFS 的存储架构以及常用 Shell 命令和 Java API，并且通过一个具体案例实现了 HDFS 的 Java API 的编程。

第 4 章主要介绍了 Hadoop 的分布式计算框架 MapReduce，分析了 MapReduce 的核心思想、工作原理、运行机制以及 MapReduce 的核心过程 Shuffle，最后通过单词计数和倒排索引两个案例详细分析 MapReduce 的编写过程和思路。

第 5 章主要介绍了 Hive 的架构、安装和相关操作，重点介绍 Hive 的 DDL、DML、DQL 操作。

第 6 章首先介绍了 HBase 的架构、寻址机制以及 HBase 的安装，然后介绍了 HBase 的 Shell 操作，包括新建表、插入数据、删除等操作，最后介绍了 HBase 常用的 Java API，并且进行了案例实现。

第 7 章主要介绍了 Spark 的基本概念和主要特点、Spark 的安装、运行架构和运行基本流程，是为学习 Spark RDD 和 Spark SQL 做基础知识储备。

第 8 章主要介绍了 RDD 的运行原理和运行流程，并对 RDD 的基本操作进行了详细的介绍，最后用一个 Scala 编程案例实现对 RDD 的操作。

第 9 章主要介绍了 Spark SQL 的原理和运行流程，并对 DataFrame 的基本操作进行了详细的介绍，最后通过三个 Scala 编程案例实现了 Spark SQL 的 DataFrame 操作、

Spark SQL 读写 MySQL 数据库和 Spark SQL 读写 Hive。

第 10 章主要介绍了 Spark Streaming 的一些基本概念和原理，介绍了 DStream 编程模型，最后通过三个 Scala 编程案例实现了 DStream 的有状态状态操作、无状态状态操作、输出操作。

第 11 章主要介绍了 Spark Streaming 与 Flume、Kafka 的整合，介绍了 Flume 和 Kafka 的安装过程，最后通过一个 Scala 编程案例实现 Spark Streaming 与 Flume、Kafka 的整合与开发。

第 12 章介绍了机器学习的定义和分类，重点介绍了 Spark MLlib 目前包含的算法和组件，通过四个具体实例 TF-IDF、线性回归、逻辑回归、协同过滤展示了利用 Spark MLlib 进行机器学习的方法和步骤。

第 13 章介绍了分布式优惠券后台应用系统的开发核心思路，优惠券后台应用系统包括商户投放子系统和用户消费子系统，分别介绍了两个子系统的核心代码以及测试调用过程。

第 14 章介绍了新闻话题实时统计分析系统的开发核心思路和核心代码以及启动调用过程。

本书特色

(1) 以实战开发为导向，对基础理论知识点与开发过程进行详细讲解。

(2) 实战案例丰富，涵盖 16 个完整项目案例和两个综合案例，综合案例可以加深对本书所学的知识点的理解和掌握。

(3) 代码详尽，避免对 API 的形式展示，规避重复代码。

(4) 语言简明易懂，由浅入深带领读者学会以 Hadoop 生态圈为核心的开发技术和大数据常见的机器学习算法。

配套资源

为便于教学，本书配有源代码、数据集、安装程序、教学课件、教学大纲。

(1) 获取源代码、数据集方式：先扫描本书封底的文泉云盘防盗码，再扫描下方二维码，即可获取。

源代码

数据集

(2) 其他配套资源可以扫描本书封底的"书圈"二维码下载。

读者对象

本书主要面向广大从事大数据分析、应用开发、机器学习、数据挖掘的专业人员以及从事高校信息技术专业的教师和高等院校的在读学生及相关领域的广大科研人员。

本书在编写过程中参考了诸多相关资料，在此对原作者表示衷心的感谢。限于作者水平和时间仓促，书中难免存在疏漏之处，欢迎读者批评指正。

作　者

2022 年 1 月

目 录

第1章 大数据概述 ··· 1
  1.1 大数据的研究背景 ··· 1
  1.2 大数据的定义及其技术特点 ··· 2
    1.2.1 大数据的定义 ··· 2
    1.2.2 大数据的基本特点 ··· 2
    1.2.3 典型的大数据处理需求与计算特征 ··························· 2
  1.3 大数据处理的主要技术特点与难点 ··································· 3
  1.4 研究大数据的意义 ··· 3
  1.5 本章小结 ·· 4

第2章 Hadoop 简介及安装部署 ·· 5
  2.1 Hadoop 简介及生态体系 ··· 5
  2.2 Hadoop 集群架构 ··· 7
  2.3 Hadoop 集群运行环境搭建 ··· 8
    2.3.1 Hadoop 安装配置过程 ·· 8
    2.3.2 验证 Hadoop 的安装 ·· 15
  2.4 本章小结 ·· 17

第3章 HDFS ··· 18
  3.1 相关基本概念 ·· 18
  3.2 HDFS 存储架构 ··· 18
    3.2.1 HDFS 写入流程 ·· 19
    3.2.2 HDFS 读取流程 ·· 19
  3.3 HDFS 的优点与缺点 ·· 20
    3.3.1 HDFS 的优点 ·· 20
    3.3.2 HDFS 的缺点 ·· 20
  3.4 HDFS Shell 常用命令 ·· 20
  3.5 HDFS 的 Java API ··· 21
  3.6 本章小结 ·· 25

第4章 MapReduce 计算框架 ·· 26
  4.1 MapReduce 核心思想 ·· 26
  4.2 MapReduce 的工作原理 ··· 28
  4.3 MapReduce 的运行机制 ··· 29
  4.4 MapReduce 数据本地化 ··· 33

4.5 MapReduce 编程 …………………………………………………………… 33
4.5.1 MapReduce 运行模式 …………………………………………… 33
4.5.2 MapReduce 编程组件与数据类型 …………………………… 34
4.6 MapReduce 编程示例 ………………………………………………… 34
4.6.1 单词计数 ………………………………………………………… 34
4.6.2 倒排索引 ………………………………………………………… 42
4.7 本章小结 ………………………………………………………………… 47

第 5 章 Hive 数据仓库 …………………………………………………………… 48
5.1 Hive 概述 ………………………………………………………………… 48
5.1.1 Hive 简介 ………………………………………………………… 48
5.1.2 Hive 的架构 ……………………………………………………… 49
5.1.3 Hive 的优缺点 …………………………………………………… 50
5.2 Hive 的安装 ……………………………………………………………… 50
5.2.1 安装 MySQL ……………………………………………………… 50
5.2.2 安装 Hive ………………………………………………………… 52
5.3 Hive 数据库相关操作 …………………………………………………… 53
5.3.1 Hive 的数据类型 ………………………………………………… 54
5.3.2 Hive 基础 SQL 语法 …………………………………………… 54
5.4 本章小结 ………………………………………………………………… 62

第 6 章 HBase 分布式数据库 …………………………………………………… 63
6.1 HBase 概述 ……………………………………………………………… 63
6.1.1 HBase 的架构 …………………………………………………… 63
6.1.2 HBase 的特点 …………………………………………………… 65
6.1.3 HBase 数据存储方式 …………………………………………… 65
6.1.4 HBase 寻址机制 ………………………………………………… 66
6.2 HBase 的安装 …………………………………………………………… 67
6.3 HBase 数据模型 ………………………………………………………… 69
6.4 HBase 的 Shell 操作 …………………………………………………… 70
6.5 HBase 常用的 Java API 及示例程序 ………………………………… 73
6.5.1 HBase 常用的 Java API ……………………………………… 73
6.5.2 程序示例 ………………………………………………………… 74
6.6 本章小结 ………………………………………………………………… 79

第 7 章 Spark 基础 ……………………………………………………………… 80
7.1 Spark 概述 ……………………………………………………………… 80
7.1.1 Spark 的主要特点 ……………………………………………… 80
7.1.2 Spark 生态系统 ………………………………………………… 81
7.1.3 Spark 相对于 Hadoop MapReduce 的优势 ………………… 82
7.2 Spark 的安装 …………………………………………………………… 82

  7.2.1 Spark 的部署方式 ·········· 82
  7.2.2 Spark 的安装 ············· 83
 7.3 Spark 运行架构与原理 ············ 85
 7.4 Spark 运行流程 ··············· 86
 7.5 本章小结 ··················· 86

第 8 章 Spark RDD 弹性分布式数据集 ······ 87
 8.1 RDD 的设计与运行原理 ············ 87
  8.1.1 RDD 的概念 ············· 87
  8.1.2 RDD 的分区 ············· 88
  8.1.3 RDD 的依赖关系 ··········· 88
  8.1.4 RDD 在 Spark 中的运行流程 ······ 90
  8.1.5 RDD 容错机制 ············ 91
 8.2 RDD API 编程 ················ 91
  8.2.1 RDD 的创建 ············· 91
  8.2.2 RDD 的操作 ············· 92
 8.3 程序示例：倒排索引 ············· 94
 8.4 本章小结 ··················· 97

第 9 章 Spark SQL ·················· 98
 9.1 Spark SQL 概述 ················ 98
  9.1.1 Spark SQL 简介 ··········· 98
  9.1.2 Spark SQL 的架构 ·········· 98
 9.2 DataFrame ··················· 99
  9.2.1 DataFrame 简介 ··········· 99
  9.2.2 DataFrame 的创建 ·········· 100
  9.2.3 DataFrame 的常用操作 ········ 101
 9.3 Dataset ···················· 102
 9.4 Spark SQL 编程 ················ 103
  9.4.1 DataFrame 操作 ··········· 103
  9.4.2 Spark SQL 读写 MySQL 数据库 ···· 106
  9.4.3 Spark SQL 读写 Hive ········· 109
 9.5 本章小结 ··················· 112

第 10 章 Spark Streaming 实时计算框架 ······ 113
 10.1 Spark Streaming 概述 ············· 113
  10.1.1 流数据和流计算 ··········· 113
  10.1.2 Spark Streaming 简介 ········ 114
  10.1.3 DStream 简介 ············ 115
 10.2 DStream 编程 ················ 115
  10.2.1 DStream 转换操作 ·········· 115

10.2.2 DStream 输出操作相关的方法 …………………… 117
10.3 DStream 编程示例 …………………… 117
   10.3.1 DStream 编程基本步骤——文件流 …………………… 117
   10.3.2 无状态转换操作 …………………… 119
   10.3.3 有状态转换操作 …………………… 122
   10.3.4 输出操作 …………………… 124
10.4 本章小结 …………………… 126

第 11 章 Spark Streaming 与 Flume、Kafka 的整合 …………………… 127
11.1 Flume 简介及安装 …………………… 127
   11.1.1 Flume 简介 …………………… 127
   11.1.2 Flume 的安装 …………………… 128
11.2 Kafka 简介及安装 …………………… 130
   11.2.1 Kafka 简介 …………………… 130
   11.2.2 Kafka 的安装 …………………… 130
11.3 Flume 与 Kafka 的区别和侧重点 …………………… 132
11.4 Spark Streaming 与 Flume、Kafka 的整合与开发 …………………… 132
11.5 本章小结 …………………… 138

第 12 章 Spark MLlib 机器学习 …………………… 139
12.1 机器学习的概念 …………………… 139
   12.1.1 机器学习的定义 …………………… 139
   12.1.2 机器学习的分类 …………………… 140
12.2 MLlib 简介 …………………… 141
12.3 Spark MLlib 的数据类型 …………………… 142
   12.3.1 本地向量 …………………… 142
   12.3.2 标注点 …………………… 142
   12.3.3 本地矩阵 …………………… 142
12.4 Spark MLlib 机器学习示例 …………………… 143
   12.4.1 特征抽取——TF-IDF …………………… 143
   12.4.2 分类与回归——线性回归 …………………… 145
   12.4.3 分类与回归——逻辑回归 …………………… 146
   12.4.4 协同过滤——电影推荐 …………………… 148
12.5 本章小结 …………………… 152

第 13 章 实战案例——分布式优惠券后台应用系统 …………………… 153
13.1 系统简介 …………………… 153
13.2 整体架构 …………………… 153
13.3 表结构设计 …………………… 154
13.4 系统实现 …………………… 155
   13.4.1 商户投放子系统 …………………… 155

  13.4.2 用户消费子系统 …… 162
 13.5 系统运行测试 …… 169
  13.5.1 启动系统 …… 169
  13.5.2 商户投放子系统测试 …… 170
  13.5.3 用户消费子系统测试 …… 172
 13.6 本章小结 …… 173

第14章 实战案例——新闻话题实时统计分析系统 …… 174
 14.1 系统简介 …… 174
 14.2 系统总体架构 …… 174
 14.3 表结构设计 …… 175
 14.4 系统实现 …… 176
  14.4.1 模拟日志生成程序 …… 176
  14.4.2 Flume 配置 …… 178
  14.4.3 配置 Kafka …… 182
  14.4.4 Spark Streaming 开发 …… 182
  14.4.5 WebSocket 和前端界面开发 …… 184
 14.5 系统运行测试 …… 192
 14.6 本章小结 …… 194

参考文献 …… 195

第1章

大数据概述

1.1 大数据的研究背景

随着计算机和信息技术的发展和普及,在人类的生产、生活中,人、资金、商品、信息的流动都以数据化方式呈现,社会正在全面迈向数字化时代,进而引发了行业数据的爆炸性增长,促进了大数据的迅猛发展。目前,大数据的应用已经渗透到了众多行业,庞大的数据资源已经成为国家和企业的战略资源。大规模的数据资源蕴涵着巨大的社会价值和商业价值,如果人们能有效地管理这些数据、挖掘数据的深度价值,则将给国家治理、社会管理、企业决策和个人生活带来巨大影响和深远的作用。因此,大数据的研究和应用已经成为全球科技创新和经济发展的重要推动力量。

大数据给人们带来新的发展机遇的同时,也带来很多新的技术挑战。例如,格式多样、形态复杂、规模庞大的行业大数据给传统的计算技术带来了很多技术困难。传统的数据库等信息处理技术已经难以应对大数据的处理需求。为此,人们亟需寻找有效的大数据处理技术方法和手段有效地处理和分析应用行业大数据。

大数据广泛且强烈的应用需求极大地推动了大数据技术的快速发展,促进了大数据处理相关基础理论方法、关键技术以及系统平台的长足发展。近十年来,在众多大数据处理技术中,出现了以 Hadoop、Spark 等为代表的主流大数据处理技术和系统平台。它们提供了基础性的大数据处理技术方法,利用大规模分布式存储和并行化计算技术,为大数据的处理和分析带来了有效的技术手段。然而,当前的大数据处理技术系统的发展尚未成熟。由于大数据的复杂性、多样性及巨大的数据规模,现有的大数据处理技术和系统平台仍然存在诸多需要不断研究和改进的技术问题,具体包括大数据的分布式存储管理技术与系统平台、并行化计算技术与系统平台、高效的并行化算法设计及易于使用的大数据分析方法与工具等。

1.2 大数据的定义及其技术特点

1.2.1 大数据的定义

对大数据很难有一个确切的定义。现有的定义都是从数据规模和支持软件处理能力角度进行的定性描述。例如,维基百科的定性描述为:大数据(Big Data)是指无法使用传统和常用的软件技术和工具在一定时间内完成获取、管理和处理的数据集。麦肯锡咨询公司的大数据报告中给出的定义是:大数据指的是在大小超出常规的数据库工具获取、存储、管理和分析能力的数据集。这些定性化的定义都无一例外地突出了大数据数据规模的"大"。实际上,如今"大数据"一词的重点已经远远超出了数据规模的定义,因为它代表着信息技术发展到了一个崭新的时代,代表着海量数据处理所需要的新技术和新方法,也代表着大数据应用所带来的新服务和新价值。

1.2.2 大数据的基本特点

相比于传统处理的数据,大数据具有如下"5V"的基本特点。

(1) 数据规模大(Volume)。即可从数百太字节(TB)到数十、数百皮字节(PB)乃至艾字节(EB)的规模。

(2) 数据多样性(Variety)。即大数据包括结构化、半结构化或非结构化等各种格式,以及数值、文本、图形、图像、流媒体等多种形态的数据。

(3) 数据处理时效性(Velocity)。即很多大数据应用需要进行及时处理,满足一定的响应性能要求。

(4) 结果准确性(Veracity)。即处理的结果要保证一定的准确性,不能因为大规模数据处理的时效性而牺牲处理结果的准确性。

(5) 深度价值(Value)。即大数据蕴含很多深度的价值,需要对大数据进行分析,挖掘出其巨大的价值。

1.2.3 典型的大数据处理需求与计算特征

大数据的"5V"基本特点给大数据的处理和利用带来了很多新的技术要求和挑战。例如,传统的数据库系统主要面向现实世界中一小部分格式较为规整的结构化数据的存储和处理。现实世界数据大多是文本和媒体等非结构化数据,这些大规模非结构化数据在传统数据处理时代大都未能得到充分的处理和利用。如何对多样化的海量数据进行实时采集、存储、分析和挖掘,让大数据产生巨大的价值是对大数据技术提出的要求和挑战。

为了有效地应对现实世界中复杂多样的大数据处理的需求,需要针对不同的大数据应用特征,总结并梳理出不同的大数据处理需求和计算特征。可以从多个角度对大数据的处理需求和计算特征进行分类。

(1) 从数据存储管理结构特征角度看,大数据可分为半结构化(或非结构化)数据和结构化数据。

(2）从数据分析类型角度看,大数据处理基本可分为传统查询分析计算和复杂数据分析挖掘处理。

（3）从数据获取处理方式角度看,大数据可分为批处理与流式计算方式。

（4）从大数据处理响应性能角度看,大数据处理也可分为实时与非实时计算,或者是联机(Online)计算与线下(Offline)计算;前述的流式计算通常属于实时计算,查询分析类计算通常也要求具有高响应性能,而批处理和复杂数据分析挖掘计算通常属于非实时或线下计算。

（5）从数据关联性角度看,大数据可分为简单关联数据(如关系表数据)和复杂关联数据(如社会网络数据)。

（6）从体系结构特征角度看,由于需要支持大规模数据的有效存储和高效计算,基于集群的大规模分布式存储与并行化计算是目前大数据处理主要采用的系统和硬件平台。

1.3 大数据处理的主要技术特点与难点

（1）技术综合性、交叉性强。大数据处理是一种涉及计算机技术众多层面的综合性计算技术。正如徐宗本院士所指出的,大数据技术需要多学科综合研究,涉及数据的获取与管理、数据的存储与处理、数据的分析与理解及结合领域的大数据应用。一个完整的大数据处理与应用系统,通常是一个包括并综合大规模硬件资源和基础设施管理、分布式存储管理、并行化计算、分析挖掘、应用服务在内的完整的技术栈。

（2）数据规模大,传统计算方法和系统失效,计算性能问题突出。大数据给传统的计算技术带来了很多新的挑战。巨大的数据量会导致巨大的计算时间开销,从而使得传统计算方法在面对大规模数据时难以在可接受的时间内完成处理。超大的数据量或计算量在计算性能方面给大数据处理技术提出了巨大挑战。

（3）应用需求驱动特性。大数据应用的很多问题都来自具体行业,大数据处理具有很强的行业应用需求驱动特性。因此,大数据处理必须紧密结合行业应用的实际场景和需求,从行业实际应用需求出发,结合实际应用需求去解决大数据处理中的技术难题,从而有效地利用大数据技术提升行业的信息处理与服务水平、发掘行业的深层价值。由于大数据技术具有典型的行业应用需求驱动特性,这也需要应用行业与计算机领域进行交叉融合。正如徐宗本院士所总结的:数据资源是基础,处理平台是支撑,分析算法是核心,应用效益是根本。

1.4 研究大数据的意义

研究大数据,最重要的意义是预测。因为数据从根本上讲是对过去和现在的归纳和总结,其本身不具备趋势和方向性的特征,但是可以应用大数据去了解事物发展的客观规律、了解人类的行为,并且能够帮助人们改变过去的思维方式,建立新的数据思维模型,从而对未来进行预测和推测。例如,谷歌公司对其用户每天频繁搜索的词汇进行数据挖掘,从而进行相关的广告推广和商业研究。

1.5 本章小结

本章主要介绍了大数据的研究背景,并对大数据的定义和基本特点以及大数据处理的主要技术特点与难点做了详细的介绍,最后阐述了研究大数据的意义。

第 2 章

Hadoop简介及安装部署

2.1 Hadoop 简介及生态体系

Hadoop 起源于 Apache Nutch 项目,始于 2002 年,是 Apache Lucene 的子项目之一。2004 年,Doug Cutting 等人受到谷歌公司在 OSDI(Operating System Design and Implementation)会议上公开发表的一篇论文的启发,开始尝试实现 MapReduce 计算框架,并将它与 NDFS(Nutch Distributed File System)结合,用作为支持 Nutch 引擎的主要算法。由于 NDFS 和 MapReduce 在 Nutch 引擎中有着良好的应用,所以它们于 2006 年 2 月被分离出来,成为一套完整而独立的软件,并被命名为 Hadoop。Hadoop 这个名字不是一个缩写,而是一个虚构的名字。该项目的创建者 Doug Cutting 这样解释 Hadoop 的命名:"Hadoop 是我孩子的一个棕黄色的大象玩具的名字。"目前,Hadoop 已成为 Apache 的顶级项目,包含众多子项目,被应用到包括雅虎公司在内的很多互联网公司。

从 Hadoop 发展到现在,Hadoop 可以大致划分为三个版本: Hadoop 1.0、Hadoop 2.0 和 Hadoop 3.0。

Hadoop 1.0 是由 Hadoop 分布式存储系统(HDFS)和分布式计算框架 MapReduce 组成。

Hadoop 2.0 在 Hadoop 1.0 的基础上,还引入了资源管理框架 YARN。

Hadoop 3.0 中引入了一些重要的功能和优化,包括 HDFS 可擦除编码、多 NameNode 支持、MapReduce 性能优化、YARN 基于 cgroup 的内存和磁盘 IO 隔离等。

Hadoop 包括两部分。一是 Hadoop 核心技术,对应为 Apache 开源社区的一个项目,主要包括 HDFS、MapReduce、YARN。其中,HDFS 用来存储海量数据; MapReduce 用来对海量数据进行计算; YARN 是一个通用的资源调度框架(在 Hadoop 2.0 中产生)。二是广义上的一个生态圈,泛指大数据技术相关的开源组件或产品,如 HBase、Hive、Spark、Pig、Kafka、Flume、Sqoop 等。生态圈中的这些组件或产品相互之间会有依赖,但又各自独立。Hadoop 生态体系如图 2-1 所示。

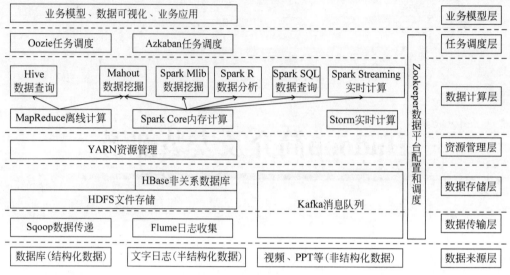

图 2-1　Hadoop 生态体系

(1) Sqoop。Sqoop 是一款开源工具,主要用于在 Hadoop、Hive 与传统的数据库(MySQL)间进行数据的传递,可以将一个关系数据库(例如 MySQL、Oracle 等)中的数据导入 Hadoop 的 HDFS 中,也可以将 HDFS 的数据导入关系数据库中。

(2) Flume。Flume 是 Cloudera 提供的一个高可用的、高可靠的、分布式的海量日志采集、聚合和传输的系统。Flume 支持在日志系统中定制各类数据发送方,用于收集数据;同时 Flume 提供对数据进行简单处理,并写入各种数据接收方的能力。

(3) Kafka。Kafka 是一种高吞吐量的分布式发布订阅消息系统,有如下特性。

① 提供消息的持久化,即使数 TB 的消息存储也能够保持长时间的稳定性能。

② 高吞吐量,可以支持每秒数百万的消息。

③ 支持通过 Kafka 服务器和消费集群来分区消息。

④ 支持 Hadoop 并行数据加载。

(4) HBase。HBase 是一个分布式的、面向列的开源数据库。HBase 不同于一般的关系数据库,它是一个适合于非结构化数据存储的数据库。

(5) HDFS。分布式存储系统,用于存储结构化或非结构化数据。

(6) YARN。资源管理和任务调度器。

(7) Storm。Storm 用于"连续计算",对数据流做连续查询,在计算时就将结果以流的形式输出给用户。

(8) MapReduce。一种分布式海量数据处理的编程模型,用于大规模数据集的并行运算。

(9) Spark。Spark 借鉴了 MapReduce 的思想并在其基础上发展起来的内存计算框架。

（10）Hive。Hive是基于Hadoop的一个数据仓库工具，可以将结构化的数据文件映射为一张数据库表，并提供简单的SQL查询功能，可以将SQL语句转换为MapReduce任务运行。其优点是学习成本低，可以通过类SQL语句快速实现简单的MapReduce统计，不必开发专门的MapReduce应用，十分适合数据仓库的统计分析。

（11）Oozie和Azkaban。Oozie和Azkaban是管理Hadoop作业的工作流程调度工具。

（12）Zookeeper。Zookeeper是一个分布式应用程序协调服务，提供的功能包括配置维护、名字服务、分布式同步、组服务等。Zookeeper的目标就是封装好复杂、易出错的关键服务，将简单易用的接口和性能高效、功能稳定的系统提供给用户。

2.2 Hadoop集群架构

Hadoop是一个大数据通用处理平台，提供了分布式文件存储以及分布式离线并行计算。由于Hadoop的高拓展性，在使用Hadoop时通常以集群的方式运行，集群中的节点可达上千个，能够处理皮字节（PB）级的数据。Hadoop集群架构如图2-2所示。

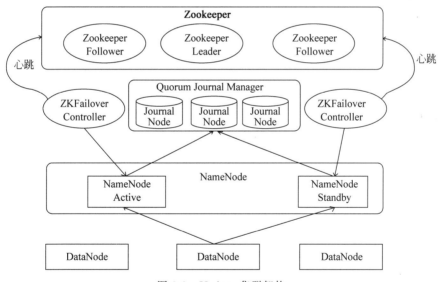

图2-2　Hadoop集群架构

Hadoop集群主要有5个部分组成。

（1）Zookeeper。它是一个分布式应用程序协调服务，主要用来解决分布式集群中应用系统的一致性问题和单点故障问题。Zookeeper有一个Leader多个Follower，基于一定的策略来保证Zookeeper的稳定性和可用性，实现分布式应用的可靠性。

（2）ZKFailover Controller（ZKFC，故障恢复控制器）。Zookeeper给每个NameNode分配一个故障恢复控制器。它负责监控NameNode的状态，并及时地把状态信息写入Zookeeper。若主NameNode发生故障，Zookeeper通知备用NameNode启动，使其成为活动状态处理客户端请求,从而实现高可用。

(3) NameNode。NameNode 通常称为名称节点或者主节点,用来保存文件系统的元数据信息(包含目录结构、数据块位置等)。在高可用 HDFS 中,通常有两台或两台以上机器充当 NameNode。无论何时,都要保证至少有一台处于活动(Active)状态,一台处于备用(Standby)状态。NameNode 主要维护两种类型文件:一种是 Fsimage,保存了最新的元数据检查点,包含了整个 HDFS 文件系统的所有目录和文件的信息,另一种是 EditLog,主要是记录 HDFS 进行的各种更新操作。

(4) DataNode。DataNode 通常称为数据节点。文件系统存储文件的方式是将文件切分成多个数据块,这些数据块实际上是存储在 DataNode 节点中的,因此 DataNode 机器需要配置大量磁盘空间。它与 NameNode 保持不断的通信,DataNode 在客户端或者 NameNode 的调度下,存储并检索数据块,对数据块进行创建、删除等操作,并且定期向 NameNode 发送所存储的数据块列表。

(5) Quorum Journal Manager(QJM)。QJM 是为了解决 NameNode 单点故障问题,Hadoop2.0 版本以后给出的高可用 HA 方案,实现 JournalNode 集群的 EditLog 操作。QJM 一般由奇数个 JournalNode 组成。每次 NameNode 写 EditLog 时,除了向本地磁盘写入 EditLog 之外,也并行地向 JournalNode 集群之中的每一个 JournalNode 发送写请求。

2.3 Hadoop 集群运行环境搭建

Hadoop 集群构建需要多台服务器,在个人学习和开发中购买多台服务器成本较大,因此,可以使用虚拟化软件在同一台计算机上构建多个 Linux 虚拟机环境,比较流行的虚拟化软件有 Virtualbox 和 VMware 等。本书采用的虚拟化软件是 VMware 的 Workstation 版本。Linux 操作系统采用的是 CentOS 7.0。

首先要使用虚拟化软件构建三台虚拟机,并在虚拟机上安装 Linux 操作系统。配置三台机器的 IP,保证三台机器之间可以相互连通(可以用 ping 命令)。Hadoop 集群配置如表 2-1 所示。

表 2-1 Hadoop 集群配置

守护进程	bigdata01 /172.16.106.69	bigdata02 /172.16.106.70	bigdata03 /172.16.106.71
NameNode	是	是	否
DataNode	是	是	是
ResourceManager	是	是	否
JournalNode	是	是	是
Zookeeper	是	是	是

2.3.1 Hadoop 安装配置过程

(1) 配置 IP。可以安装 Linux 图形界面,在图形界面上进行配置;也可以安装

Xshell 等客户端软件连接到 Linux 虚拟机上进行命令行下配置(如果对命令行不熟的学习者,建议用图形界面)。

(2) 修改主机名。

① 在/etc/sysconfig/network 文件中添加以下内容(例如在第一台机器上配置):

```
NETWORKING = yes
HOSTNAME = bigdata01
```

② 在/etc/hosts 文件中添加主机名和 IP 的映射关系,配置如下:

```
172.16.106.69    bigdata01
172.16.106.70    bigdata02
172.16.1068.71   bigdata03
```

③ 在/etc/hostname 文件中添加机器名称,配置如下:

```
bigdata01
```

(3) 关闭防火墙,在 CentOS 命令行输入如下命令:

```
systemctl stop firewalld
systemctl disable firewalld
```

(4) 重启操作系统。在 CentOS 命令行输入 reboot 命令,重启系统。

(5) 检查配置是否正确。

在 CentOS 命令行输入 hostname 命令检查主机名是否配置正确,用 ping bigdata01 命令查看机器是否连通。成功后将另外两台机器做同样的配置,主机名分别为 bigdata02 和 bigdata03,IP 也做相应的修改配置。

(6) 上传安装包(可以使用 Xftp 上传 JDK、Hadoop、Zookeeper 安装文件,本书使用的 JDK 版本是 1.8,Hadoop 版本是 2.7.7,Zookeeper 版本是 3.4.13)。

(7) 建立 JDK、Hadoop 安装目录,在 CentOS 命令行输入如下命令:

```
cd /
mkdir /usr/java
mkdir hadoop
```

(8) 配置 JDK,将 Hadoop 添加到环境变量中。

① 解压安装包,在 CentOS 命令行输入如下命令:

```
tar -zxvf jdk-8u161-linux-x64.tar.gz -C /usr/java
```

② 在/etc/profile 文件中添加配置如下:

```
export JAVA_HOME = /usr/java/jdk1.8.0_161
export HADOOP_HOME = /hadoop/hadoop-2.7.7
exportPATH = $PATH:$JAVA_HOME/bin:$HADOOP_HOME/bin:$HADOOP_HOME/sbin
```

保存后,在 CentOS 命令行中输入 source /etc/profile 命令使配置生效。

注意,以上操作在 3 台机器上分别进行。

(9) 安装配置 Zookeeper 集群(在 bigdata01 上)。

① 解压 Zookeeper 安装包,在 CentOS 命令行输入如下命令:

```
tar -zxvf zookeeper-3.4.13.tar.gz -C /hadoop/
```

② 配置 zoo.cfg 文件。

首先进入 conf 目录,在 CentOS 命令行中输入命令 cp zoo_sample.cfg zoo.cfg,生成 zoo.cfg 文件,然后修改该文件,添加配置如下:

```
dataDir = /hadoop/zookeeper-3.4.13/tmp
server.1 = bigdata01:2888:3888
server.2 = bigdata02:2888:3888
server.3 = bigdata03:2888:3888
```

创建一个 tmp 文件夹,在 CentOS 命令行中输入命令 mkdir /hadoop/zookeeper-3.4.13/tmp,然后创建一个空文件,向该文件写入 ID,在 CentOS 命令行中输入命令:

```
echo 1 > /hadoop/zookeeper-3.4.13/tmp/myid
```

③ 将配置好的 Zookeeper 复制到其他节点(首先分别在 bigdata02、bigdata03 根目录下创建一个 hadoop 目录),在 CentOS 命令行中输入如下命令:

```
scp -r /hadoop/zookeeper-3.4.13/  root@bigdata02:/hadoop/
scp -r /hadoop/zookeeper-3.4.13/  root@bigdata03:/hadoop/
```

④ 修改 bigdata02、bigdata03 对应 /hadoop/zookeeper-3.4.13/tmp/myid 内容。

在 bigdata02 机器上,在 CentOS 命令行中输入如下命令:

```
echo 2 > /hadoop/zookeeper-3.4.13/tmp/myid
```

在 bigdata03 机器上,在 CentOS 命令行中输入如下命令:

```
echo 3 > /hadoop/zookeeper-3.4.13/tmp/myid
```

(10) 安装配置 Hadoop 集群(在 bigdata01 上操作)。

① 解压 Hadoop 安装包,在 CentOS 命令行中输入如下命令:

```
tar -zxvf hadoop-2.7.7.tar.gz -C /hadoop/
```

② 配置 HDFS,Hadoop 2.0 所有的配置文件都在 $HADOOP_HOME/etc/hadoop 目录下。进入配置目录,在 CentOS 命令行中输入如下命令:

```
cd /hadoop/hadoop-2.7.7/etc/hadoop
```

③ 修改 hadoo-env.sh 文件,配置如下:

```
export JAVA_HOME = /usr/java/jdk1.8.0_161
```

④ 修改 core-site.xml 文件,配置如下:

```
<configuration>
  <!-- 指定 HDFS 的 NameService 为 ns1 -->
```

```xml
<property>
<name>fs.defaultFS</name>
<value>hdfs://ns1</value>
</property>
<!-- 指定 Hadoop 临时目录 -->
<property>
<name>hadoop.tmp.dir</name>
<value>/hadoop/hadoop-2.7.7/tmp</value>
</property>
<!-- 指定 Zookeeper 地址 -->
<property>
<name>ha.zookeeper.quorum</name>
<value>bigdata01:2181,bigdata02:2181,bigdata03:2181</value>
</property>
</configuration>
```

⑤ 修改 hdfs-site.xml 文件，配置如下：

```xml
<configuration>
<!-- 指定 HDFS 的 NameService 为 ns1,需要和 core-site.xml 中的保持一致 -->
<property>
<name>dfs.nameservices</name>
<value>ns1</value>
</property>
<!-- ns1 下面有两个 NameNode,分别是 nn1,nn2 -->
<property>
<name>dfs.ha.namenodes.ns1</name>
<value>nn1,nn2</value>
</property>
<!-- nn1 的 RPC 通信地址 -->
<property>
<name>dfs.namenode.rpc-address.ns1.nn1</name>
<value>bigdata01:9000</value>
</property>
<!-- nn1 的 HTTP 通信地址 -->
<property>
<name>dfs.namenode.http-address.ns1.nn1</name>
<value>bigdata01:50070</value>
</property>
<!-- nn2 的 RPC 通信地址 -->
<property>
<name>dfs.namenode.rpc-address.ns1.nn2</name>
<value>bigdata02:9000</value>
</property>
<!-- nn2 的 HTTP 通信地址 -->
<property>
<name>dfs.namenode.http-address.ns1.nn2</name>
<value>bigdata02:50070</value>
</property>
<!-- 指定 NameNode 的元数据在 JournalNode 上的存放位置 -->
```

```xml
<property>
<name>dfs.namenode.shared.edits.dir</name>
<value>qjournal://bigdata01:8485;bigdata02:8485;bigdata03:8485/ns1</value>
</property>
<!-- 指定JournalNode在本地磁盘存放数据的位置 -->
<property>
<name>dfs.journalnode.edits.dir</name>
<value>/hadoop/hadoop-2.7.7/journal</value>
</property>
<!-- 开启NameNode失败自动切换 -->
<property>
<name>dfs.ha.automatic-failover.enabled</name>
<value>true</value>
</property>
<!-- 配置失败自动切换实现方式 -->
<property>
<name>dfs.client.failover.proxy.provider.ns1</name>
<value>org.apache.hadoop.hdfs.server.namenode.ha.ConfiguredFailoverProxyProvider</value>
</property>
<!-- 配置隔离机制方法,多个机制用换行分隔,即每个机制暂用一行 -->
<property>
<name>dfs.ha.fencing.methods</name>
<value>
sshfence
Shell(/bin/true)
</value>
</property>
<!-- 使用sshfence隔离机制时需要SSH免登录 -->
<property>
<name>dfs.ha.fencing.ssh.private-key-files</name>
<value>/root/.ssh/id_rsa</value>
</property>
<!-- 配置sshfence隔离机制超时时间 -->
<property>
<name>dfs.ha.fencing.ssh.connect-timeout</name>
<value>30000</value>
</property>
<property>
<name>dfs.replication</name>
<value>1</value>
</property>
</configuration>
```

⑥ 修改Mapred-site.xml文件,配置如下:

```xml
<configuration>
<!-- 指定MR框架为YARN方式 -->
<property>
<name>MapReduce.framework.name</name>
<value>yarn</value>
```

```
</property>
</configuration>
```

此处注意,需要生成 Mapred-site.xml 文件,命令如下:

```
cp Mapred-site.xml.template Mapred-site.xml
```

⑦ 修改 yarn-site.xml 文件,配置如下:

```
<configuration>
  <!-- 开启 RM 高可靠 -->
  <property>
      <name>yarn.resourcemanager.ha.enabled</name>
      <value>true</value>
  </property>
  <!-- 指定 RM 的 cluster id -->
  <property>
      <name>yarn.resourcemanager.cluster-id</name>
      <value>yrc</value>
  </property>
  <!-- 指定 RM 的名字 -->
  <property>
      <name>yarn.resourcemanager.ha.rm-ids</name>
      <value>rm1,rm2</value>
  </property>
  <!-- 分别指定 RM 的地址 -->
  <property>
      <name>yarn.resourcemanager.hostname.rm1</name>
      <value>bigdata01</value>
  </property>
  <property>
      <name>yarn.resourcemanager.hostname.rm2</name>
      <value>bigdata02</value>
  </property>
  <!-- 指定 Zookeeper 集群地址 -->
  <property>
      <name>yarn.resourcemanager.zk-address</name>
      <value>bigdata01:2181,bigdata02:2181,bigdata03:2181</value>
  </property>
  <property>
      <name>yarn.nodemanager.aux-services</name>
      <value>MapReduce_shuffle</value>
  </property>
</configuration>
```

⑧ 修改 slaves 文件,配置如下:

```
bigdata01
bigdata02
bigdata03
```

(11) 配置免密码登录。

① 配置 bigdata01 到 bigdata02、bigdata03 的免密码登录。

在 bigdata01 上生产一对钥匙,在 CentOS 命令行中输入如下命令:

ssh－keygen－trsa

将公钥复制到其他节点,包括自己,在 CentOS 命令行中输入如下命令:

ssh－copy－id bigdata01
ssh－coyp－id bigdata02
ssh－coyp－id bigdata03

② 配置 bigdata02 到 bigdata01、bigdata03 的免密码登录。

在 bigdata02 上生产一对钥匙,在 CentOS 命令行中输入如下命令:

ssh－keygen－trsa

将公钥复制到其他节点,在 CentOS 命令行中输入如下命令:

ssh－copy－id bigdata01
ssh－coyp－id bigdata02
ssh－coyp－id bigdata03

注意:两个 NameNode 之间要配置 SSH 免密码登录,别忘了配置 bigdata02 到 bigdata01 的免密码登录。

③ 将配置好的 Hadoop 复制到其他节点,在 CentOS 命令行中输入如下命令:

scp－r/hadoop/hadoop－2.7.7/ root@bigdata02:/hadoop/
scp－r/hadoop/hadoop－2.7.7/ root@bigdata03:/hadoop/

(12) 启动 Hadoop。

① 启动 Zookeeper 集群(分别在 bigdata01、bigdata02、bigdata03 启动),在 CentOS 命令行中输入如下命令:

cd/hadoop/zookeeper－3.4.13/bin/
./zkServer.sh start

查看 Zookeeper 集群状态,在 CentOS 命令行中输入如下命令:

./zkServer.sh status

结果应为一个 Leader 和两个 Follower。

② 启动 JournalNode(分别在 bigdata01、bigdata02、bigdata03 上执行),在 CentOS 命令行中输入如下命令:

cd/hadoop/hadoop－2.7.7
sbin/hadoop－daemon.sh start journalnode

③ 格式化 HDFS。

在 bigdata01 上执行命令:

```
hdfs namenode - format
```

执行格式化后 Hadoop 根据 core-site.xml 文件中的 hadoop.tmp.dir 配置项生成文件,此处配置的是/hadoop/hadoop-2.7.7/tmp,然后将/hadoop/hadoop-2.7.7/tmp 复制到 bigdata02 和 bigdata03 的/hadoop/hadoop-2.7.7/下。

④ 格式化 Zookeeper(在 bigdata01 上执行即可),在 CentOS 命令行中输入如下命令:

```
hdfs zkfc - formatZK
```

⑤ 启动 HDFS(在 bigdata01 上执行),在 CentOS 命令行中输入如下命令:

```
sbin/start - dfs.sh
```

⑥ 启动 YARN,在 CentOS 命令行中输入如下命令:

```
sbin/start - yarn.sh
```

2.3.2 验证 Hadoop 的安装

在 Hadoop 集群搭建完成后,Hadoop 提供一个 Web UI 管理页面,可以观察到 Hadoop 集群的基本状况,通过浏览器访问 http://172.16.106.69:50070,主 NameNode 监控界面如图 2-3 所示。

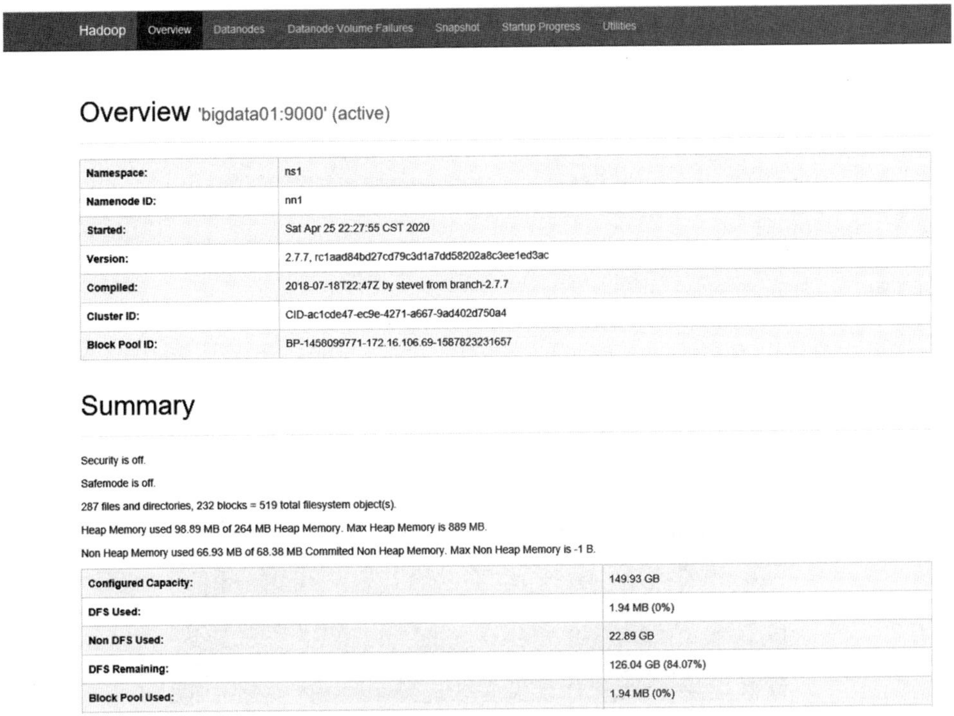

图 2-3 主 NameNode 监控界面

通过浏览器访问 http://172.16.106.70:50070，备用 NameNode 监控界面如图 2-4 所示。

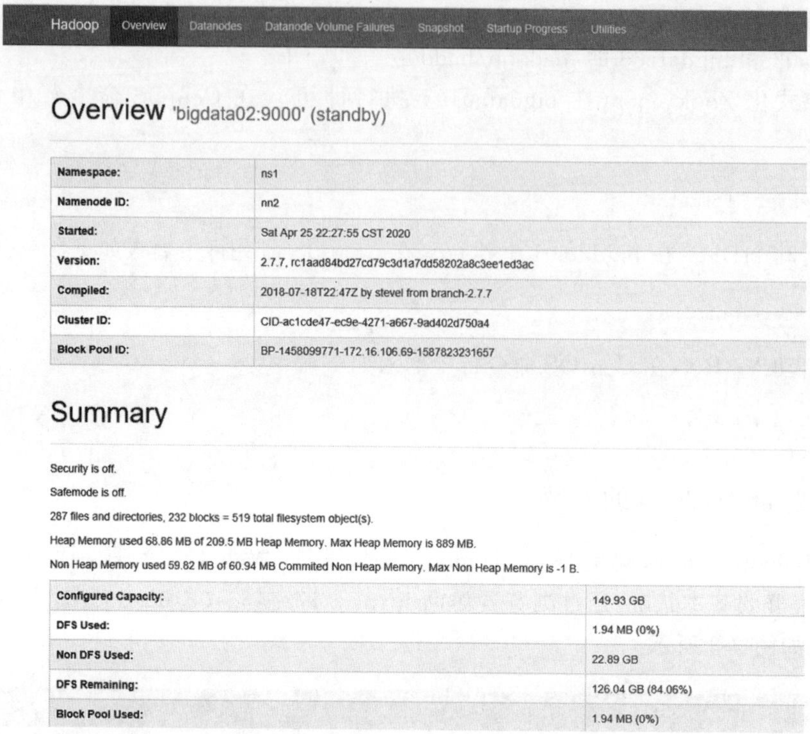

图 2-4　备用 NameNode 监控界面

从两个界面上可以看出，NameNode 一个是激活状态（Active），一个是备用状态（Standby）。通过向 HDFS 上传文件，测试 Hadoop 是否安装成功，在 CentOS 命令行中输入如下 Hadoop 命令：

```
hadoop fs -put /etc/profile /profile
hadoop fs -ls/
```

HDFS 上传文件结果如图 2-5 所示。

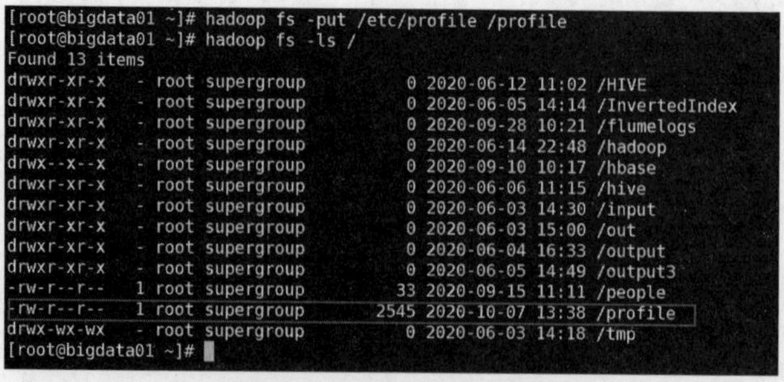

图 2-5　HDFS 上传文件结果

2.4 本章小结

本章主要介绍了 Hadoop 的起源、生态体系和集群架构,对 Hadoop 的安装步骤进行了详细的介绍,最后验证了 Hadooop 的安装。根据本章介绍的 Hadoop 安装步骤,相信读者能够搭建起自己的 Hadoop 集群。

第 3 章

HDFS

3.1 相关基本概念

(1) 文件系统。文件系统是操作系统提供的用于解决"如何在磁盘上组织文件"的一系列方法和数据结构。

(2) 分布式文件系统。分布式文件系统是指利用多台计算机协同作用解决单台计算机所不能解决的存储问题的文件系统。如单机负载高、数据不安全等问题。

(3) HDFS(Hadoop Distributed File System,Hadoop 分布式文件系统)。它是 Hadoop 项目的核心子项目,是分布式计算中数据存储管理的基础,它是基于流式数据访问和处理超大文件的需求而开发的分布式文件系统,可以运行于廉价的商用服务器上。HDFS 源于谷歌公司在 2003 年 10 月份发表的 GFS(Google File System)论文。

3.2 HDFS 存储架构

HDFS 采用 Master/Slave 架构。一个 HDFS 集群是由一个或几个 NameNode 和一定数量的 DataNode 组成的。HDFS 上的文件是以数据块的形式存放的,这些数据块存储在一组 DataNode 上。NameNode 执行文件系统的命名空间操作,比如打开、关闭、重命名文件或目录; 也负责确定数据块到具体 DataNode 节点的映射。DataNode 负责处理文件系统客户端的读写请求,并在 NameNode 的统一调度下执行数据块的创建、删除和复制。HDFS 存储架构如图 3-1 所示。

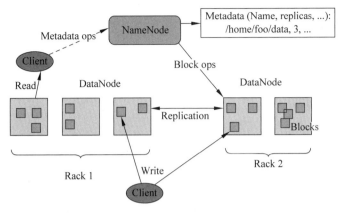

图 3-1　HDFS 存储架构

3.2.1　HDFS 写入流程

（1）Hadoop 客户端和 NameNode 通信请求上传文件，NameNode 检查目标文件是否已存在、父目录是否存在。

（2）NameNode 返回信息给 Hadoop 客户端是否可以上传。

（3）Hadoop 客户端会先对文件进行切分，如一个 Block（块）大小为 128MB，如果上传文件为 300MB，文件会被切分成 3 个 Block，其中两个 Block 为 128MB，一个 Block 为 44MB，并向 NameNode 发上传请求。

（4）NameNode 返回 DataNode 的服务器信息给 Hadoop 客户端。

（5）Hadoop 客户端请求一台 DataNode 上传数据（本质上是一个 RPC 调用，建立通道），第一个 DataNode 收到请求会继续调用第二个 DataNode，然后第二个 DataNode 调用第三个 DataNode，将整个通道建立完成，逐级返回 Hadoop 客户端。

（6）Hadoop 客户端开始往第一个 DataNode 上传第一 Block（先从磁盘读取数据放到一个本地内存缓存），以 Packet 为单位（一个 Packet 为 64KB），当然在写入时通道会进行数据校验，它并不是通过一个 Packet 进行一次校验而是以 checksum 为单位进行校验（512B），第一台 DataNode 收到一个 Packet 就会传给第二台，第二台传给第三台；第一台每传一个 Packet 会放入一个应答队列等待应答。

（7）当一个 Block 传输完成之后，Hadoop 客户端再次请求 NameNode 上传第二个 Block 的 DataNode 服务器，直至所有的 Block 上传完成。

3.2.2　HDFS 读取流程

（1）Hadoop 客户端向 NameNode 发送请求，获得存放在 NameNode 节点上文件的 Block 位置映射信息。

（2）NameNode 把文件所有 Block 的位置信息返回给 Hadoop 客户端。

（3）Hadoop 客户端拿到 Block 的位置信息后并行读取 Block 信息，Block 默认有 3 个副本，所以每一个 Block 只需要从一个副本读取。

(4) Hadoop 客户端从 DataNode 上取回文件的所有 Block 按照一定的顺序组成最终需要的文件。

3.3 HDFS 的优点与缺点

3.3.1 HDFS 的优点

(1) 高容错性。数据自动保存多个副本。它通过增加副本的形式提高容错性。某一个副本丢失以后,它可以自动恢复。

(2) 适合处理高吞吐量。HDFS 通过移动计算而不是移动数据,并将数据位置暴露给计算框架,对计算的时延不敏感。

(3) 适合存储和管理大规模数据。HDFS 处理数据达到太字节(TB)甚至皮字节(PB)级别的数据,文件数量达百万规模以上,节点处理可达 1 万节点的规模。

(4) 适合一次写入,多次读取。文件一旦写入不能修改,只能追加,能够保证数据的一致性。

(5) 适合处理非结构化数据。HDFS 可以处理多类型的数据(音频、视频、文本)。

3.3.2 HDFS 的缺点

(1) 不适合低延时数据访问。HDFS 适合高吞吐率的场景,就是在某一段时间内写入大量的数据。但是 HDFS 不适合低延时的数据访问,比如毫秒级以内读取数据是很难做到的。

(2) 不适合小文件存储。小文件是指小于 HDFS 系统的 Block 大小的文件(1.0 版本默认为 64MB,2.0 版本默认为 128MB)的文件,小文件存储会占用 NameNode 大量的内存来存储文件、目录和块信息。而 NameNode 的内存是有限的。

(3) 不支持文件随机修改。HDFS 不允许多个线程同时写文件;仅支持数据追加,不支持文件的随机修改。

3.4 HDFS Shell 常用命令

HDFS Shell 可以直接与 Hadoop 分布式文件系统进行交互。HDFS Shell 常用命令如表 3-1 所示。

表 3-1 HDFS Shell 常用命令

命令参数	功能描述	命令参数	功能描述
-ls	查看指定路径的目录结构	-text	源文件输出为文本格式
-du	统计目录下所有文件大小	-mkdir	创建空白文件夹
-mv	移动文件	-put	上传文件
-cp	复制文件	-get	下载文件
-rm	删除文件/空白文件夹	-help	获得帮助信息
-cat	查看文件内容	-cat	查看文件内容

例如，列出文件或文件夹 Shell 命令为 hadoop fs -ls/，运行结果如下：

```
[root@bigdata01 /]# hadoop fs -ls /
Found 13 items
drwxr-xr-x   - root supergroup          0 2020-06-12 11:02 /HIVE
drwxr-xr-x   - root supergroup          0 2020-06-05 14:14 /InvertedIndex
drwxr-xr-x   - root supergroup          0 2020-09-28 10:21 /flumelogs
drwxr-xr-x   - root supergroup          0 2020-06-14 22:48 /hadoop
drwx--x--x   - root supergroup          0 2020-09-10 10:17 /hbase
drwxr-xr-x   - root supergroup          0 2020-06-06 11:15 /hive
drwxr-xr-x   - root supergroup          0 2020-06-03 14:30 /input
drwxr-xr-x   - root supergroup          0 2020-06-03 15:00 /out
drwxr-xr-x   - root supergroup          0 2020-06-04 16:33 /output
drwxr-xr-x   - root supergroup          0 2020-06-05 14:49 /output3
-rw-r--r--   1 root supergroup         33 2020-09-15 11:11 /people
-rw-r--r--   1 root supergroup       2545 2020-10-07 13:38 /profile
drwx-wx-wx   - root supergroup          0 2020-06-03 14:18 /tmp
```

创建文件夹 Shell 命令为 hadoop fs -mkdir/wcdoc，运行结果如下：

```
[root@bigdata01 /]# hadoop fs -mkdir /wcdoc
[root@bigdata01 /]# hadoop fs -ls /
Found 15 items
drwxr-xr-x   - root supergroup          0 2020-06-12 11:02 /HIVE
drwxr-xr-x   - root supergroup          0 2020-06-05 14:14 /InvertedInde
drwxr-xr-x   - root supergroup          0 2020-09-28 10:21 /flumelogs
drwxr-xr-x   - root supergroup          0 2020-06-14 22:48 /hadoop
drwx--x--x   - root supergroup          0 2020-09-10 10:17 /hbase
drwxr-xr-x   - root supergroup          0 2020-06-06 11:15 /hive
drwxr-xr-x   - root supergroup          0 2020-06-03 14:30 /input
drwxr-xr-x   - root supergroup          0 2020-06-03 15:00 /out
drwxr-xr-x   - root supergroup          0 2020-06-04 16:33 /output
drwxr-xr-x   - root supergroup          0 2020-06-05 14:49 /output3
-rw-r--r--   1 root supergroup         33 2020-09-15 11:11 /people
-rw-r--r--   1 root supergroup       2545 2020-10-07 13:38 /profile
drwx-wx-wx   - root supergroup          0 2020-06-03 14:18 /tmp
drwxr-xr-x   - root supergroup          0 2020-10-08 13:13 /user
drwxr-xr-x   - root supergroup          0 2020-10-08 13:14 /wcdoc
```

上传文件 Shell 命令为 hadoop fs -put/wc.txt/wcdoc，运行结果如下：

```
[root@bigdata01 /]# hadoop fs -put /wc.txt /wcdoc
[root@bigdata01 /]# hadoop fs -ls /wcdoc
Found 4 items
drwxr-xr-x   - root supergroup          0 2020-10-08 13:24 /wcdoc/bin
drwxr-xr-x   - root supergroup          0 2020-10-08 13:24 /wcdoc/boot
drwxr-xr-x   - root supergroup          0 2020-10-08 13:24 /wcdoc/dev
-rw-r--r--   1 root supergroup         53 2020-10-08 13:25 /wcdoc/wc.txt
```

下载文件 Shell 命令为 hadoop fs -get/wcdoc/wc.txt/wc1.txt，运行结果如下：

```
[root@bigdata01 /]# hadoop fs -get /wcdoc/wc.txt /wc1.txt
[root@bigdata01 /]# ls
bin    etc       file3.txt  lib    mnt     people.json  run   sys    usr      wc.txt
      file1.txt  hadoop     lib64  opt     proc         sbin  tmp    var      words.txt
dev   file2.txt  home       media  people  root         srv   tools  wc1.txt
```

查看文件内容 Shell 命令为 hadoop fs -cat/wcdoc/wc.txt，运行结果如下：

```
[root@bigdata01 /]# hadoop fs -cat /wcdoc/wc.txt
hello hadoop
hello mapreduce
bye hadoop
```

删除文件（夹）Shell 命令为 hadoop fs -rm/wcdoc/wc.txt，运行结果如下：

```
[root@bigdata01 /]# hadoop fs -rm /wcdoc/wc.txt
20/10/08 13:29:28 INFO fs.TrashPolicyDefault: Namenode trash configuration: Deletion interval = 0 minutes, Empt
ier interval = 0 minutes.
Deleted /wcdoc/wc.txt
```

3.5 HDFS 的 Java API

通过编程的形式操作 HDFS，其核心是使用 HDFS 提供的 Java API 构造一个访问客户端对象，然后通过客户端对象对 HDFS 上的文件进行操作（增加、删除、查找）。

在 Java 中操作 HDFS,创建一个客户端实例主要涉及以下两个类:一是 Configuration 类,该类的对象封装了客户端或者服务器的配置,从中获取 Hadoop 集群的配置信息;二是 FileSystem 类,该类的对象是一个文件系统对象。

FileSystem 对象的一些方法可以对文件进行操作,HDFS 常用 Java API 如表 3-2 所示。

表 3-2　HDFS 常用 Java API

方 法 名	功　　能
copyFromLocalFile(Path src,Path dst)	从本地磁盘复制文件到 HDFS
copyToLocalFile(Path src,Path dst)	从 HDFS 复制文件到本地磁盘
mkdirs(Path f)	建立子目录
rename(Path src,Path dst)	重命名文件或文件夹
delete(Path f)	删除指定文件

搭建测试项目具体步骤如下所述。

(1) 创建一个 MAVEN 项目,并在项目的 pom.xml 文件中引入 hadoop-common、hadoop-hdfs、hadoop-client 以及单元测试 junit 的依赖,pom.xml 文件中的依赖包如下:

```xml
<dependency>
    <groupId>org.apache.hadoop</groupId>
    <artifactId>hadoop-common</artifactId>
    <version>2.7.4</version>
</dependency>
<dependency>
    <groupId>org.apache.hadoop</groupId>
    <artifactId>hadoop-hdfs</artifactId>
    <version>2.7.4</version>
</dependency>
<dependency>
    <groupId>org.apache.hadoop</groupId>
    <artifactId>hadoop-client</artifactId>
    <version>2.7.4</version>
</dependency>
<dependency>
    <groupId>org.apache.hadoop</groupId>
    <artifactId>hadoop-MapReduce-client-core</artifactId>
    <version>2.7.4</version>
</dependency>
<dependency>
    <groupId>junit</groupId>
    <artifactId>junit</artifactId>
    <version>4.12</version>
</dependency>
<dependency>
    <groupId>org.apache.zookeeper</groupId>
    <artifactId>zookeeper</artifactId>
```

```
<version>3.4.10</version>
</dependency>
```

（2）编写 Java 类 HDFS_API_TEST，代码和相关说明如下：

```java
package com.chapter03.hdfsdemo;
import java.io.FileNotFoundException;
import java.io.IOException;
import org.apache.hadoop.conf.Configuration;
import org.apache.hadoop.fs.BlockLocation;
import org.apache.hadoop.fs.FileStatus;
import org.apache.hadoop.fs.FileSystem;
import org.apache.hadoop.fs.LocatedFileStatus;
import org.apache.hadoop.fs.Path;
import org.apache.hadoop.fs.RemoteIterator;
import org.junit.Before;
import org.junit.Test;
public class HDFS_API_TEST{
    FileSystem fs = null;
    @Before
    public void init()throws Exception{
        //构造配置参数对象
        Configuration conf = new Configuration();
        //设置访问的 HDFS 的 URI
        conf.set("fs.defaultFS","hdfs://172.16.106.69:9000");
        //设置本机的 Hadoop 路径
        System.setProperty("hadoop.home.dir","D:\\hadoop");
        //设置客户端访问身份
        System.setProperty("HADOOP_USER_NAME","root");
        //通过 FileSystem 的静态 get()方法获取文件系统客户端对象
        fs = FileSystem.get(conf);
    }
    @Test
    public void AddFileToHdfs()throws IOException{
        //上传文件本地路径
        Path src = new Path("D:/hdfs.txt");
        //HDFS 的目标路径
        Path dst = new Path("/teshdfs");
        //上传文件方法
        fs.copyFromLocalFile(src,dst);
        //关闭资源
        fs.close();
    }
    //从 HDFS 中复制文件到本地文件系统
    @Test
        public void DownloadFileToLocal()throws IllegalArgumentException,IOException {
        //下载文件
        fs.copyToLocalFile(new Path("/testFile"),new Path("D:/"));
    }
    //创建、删除、重命名文件
```

```java
@Test
public void MkdirAndDeleteAndRename()throws Exception{
    //创建目录
    fs.mkdirs(new Path("/test1"));
    fs.rename(new Path("/test1"),new Path("/test3"));
    //删除文件夹,如果是非空文件夹,则参数2必须赋值true
    fs.delete(new Path("/test2"), true);
}
//查看目录信息
@Test
public void ListFiles()throws FileNotFoundException, IllegalArgumentException, IOException {
    //获取迭代器对象
    RemoteIterator<LocatedFileStatus> listFiles = fs.listFiles(new Path("/"), true);
    while (listFiles.hasNext()){
        LocatedFileStatus fileStatus = listFiles.next();
        //打印当前文件名
        System.out.println(fileStatus.getPath().getName());
        //打印当前文件块大小
        System.out.println(fileStatus.getBlockSize());
        //打印当前文件权限
        System.out.println(fileStatus.getPermission());
        //打印当前文件内容长度
        System.out.println(fileStatus.getLen());
        //获取该文件块信息(包含长度、数据块、DataNode的信息)
        BlockLocation[] blockLocations = fileStatus.getBlockLocations();
        for (BlockLocation bl : blockLocations){
            System.out.println("block-length:" + bl.getLength() + "--" + "block-offset:" + bl.getOffset());
            String[] hosts = bl.getHosts();
            for (String host : hosts){
                System.out.println(host);
            }
        }
    }
}
//查看文件及文件夹信息
@Test
public void ListFileAll()throws FileNotFoundException,IllegalArgumentException, IOException {
    //获取HDFS系统根目录的元数据信息
    FileStatus[] listStatus = fs.listStatus(new Path("/"));
    String filelog = "文件夹--     ";
    for (FileStatus fstatus : listStatus){
        //判断是文件还是文件夹
        if (fstatus.isFile())
            filelog = "文件--       ";
        System.out.println(filelog + fstatus.getPath().getName());
    }
}
```

（3）运行测试。

对于 Windows 机器上没有 Hadoop 的环境的客户端，需要下载 Hadoop 的服务进程 winutils.exe，下载地址为 https://codeload.github.com/srccodes/hadoop-common-2.2.0-bin/zip/master。解压后放在一个没有中文的文件夹下。在程序中指定路径，本例把它放在 D 盘 hadoop 目录下，同时在类的 init()方法中加入语句：System.setProperty("hadoop.home.dir", "D:\\hadoop")。指定客户端 Hadoop 的伪安装路径后，可以利用单元测试 junit 进行各个方法的运行测试。

3.6 本章小结

本章主要介绍了 Hadoop 中非常重要的分布式存储文件系统 HDFS，分析了 HDFS 的存储架构以及常用 Shell 命令和 Java API，并且对 Java API 的编程进行了实例演示。

第 4 章

MapReduce计算框架

MapReduce是一种分布式的计算模型，是 Hadoop 系统的核心组件之一，用于解决海量数据的计算问题。MapReduce 计算框架源自谷歌公司的 MapReduce 论文，Hadoop MapReduce 是谷歌公司 MapReduce 的克隆版。MapReduce 将整个并行计算过程抽象到两个函数：Map()函数和 Reduce()函数。一个简单的 MapReduce 程序只需要指定 Map、Reduce、input、format，剩下的由框架来完成。在进行数据的传输时采用的是(key,value)键值对的形式进行传输。基于 MapReduce 计算模型编写分布式并行程序非常简单，开发人员的主要编码工作就是实现 Map()函数和 Reduce()函数。其他的并行编程中的种种复杂问题，如分布式存储、工作调度、负载平衡、容错处理、网络通信等，均由 YARN 框架负责处理。

4.1 MapReduce 核心思想

（1）并行处理，分而治之。

MapReduce 的核心思想是"分而治之"。所谓"分而治之"就是把一个复杂的问题，按照一定的"分解"方法分为等价的规模较小的若干部分，然后逐个解决，分别找出各部分的结果，把各部分的结果组成整个问题的结果。MapReduce 任务分解示意图如图 4-1 所示。

任务分解的前提是这些任务没有必然的依赖关系，可以单独执行任务，将结果合并，即把任务划分中的各个子任务的结果进行全局汇总。例如，一批文本文件有 40GB。假设要统计每个单词出现的总次数。为了提高工作效率，在多个机器上进行处理。首先，将数据分成多个，然后编写一个程序同时在多个机器上运行。这样，每个机器只能统计自己机器上的数据，但是无法得出全局的结论。所以，还需要做进一步的处理，即读入每个机器的统计结果，然后进行汇总处理。

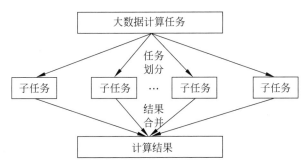

图 4-1 MapReduce 任务分解示意图

(2) 上升到编程模型：Map 与 Reduce。

MapReduce 将数据的处理方式抽象为 Map 和 Reduce，用户只需实现 Mapper 和 Reducer 这两个抽象类，编写 Map() 和 Reducer() 两个函数，即可完成简单的分布式程序的开发。Map 是过滤和聚集数据，表现为数据的一对一的映射，通常完成数据转换的工作。Reduce 是根据 Map 的生成完成归约、分组和总结，表现为多对一的映射，通常完成数据的聚合操作。

WordCount(单词统计)是最简单也是最能体现 MapReduce 思想的程序之一，可以称为 MapReduce 版 Hello World。WordCount 主要完成的功能是：统计一系列文本文件中每个单词出现的次数。MapReduce 示意图如图 4-2 所示。

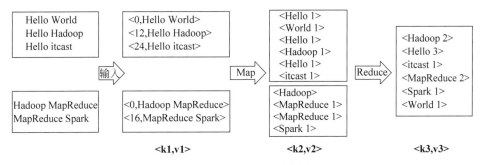

图 4-2 MapReduce 示意图

首先，MapReduce 通过默认组件 TextInputFormat 将待处理的数据文件的每一行的数据都转变为 <key,value> 键值对。其次，调用 Map() 方法，将单词进行切割并进行计数，输出键值对作为 Reduce 阶段的输入键值对。最后，调用 Reduce() 方法将单词汇总、排序后，通过 TextOutputFormat 组件输出到结果文件中。

(3) 上升到构架：以统一构架为开发人员隐藏系统层细节。

MapReduce 提供了统一的计算框架，为开发人员隐藏了绝大多数系统层面的处理细节，程序员只需要集中于应用问题和算法本身，而不需要关注其他系统层的处理细节，大大减轻了开发人员开发程序的负担。

该框架可负责自动完成以下系统底层相关的处理。

① 计算任务的自动划分和调度。

② 数据的自动化分布存储和划分。
③ 处理数据与计算任务的同步。
④ 结果数据的收集整理(排序、合并和分区等)。
⑤ 系统通信、负载平衡、计算性能优化处理。
⑥ 处理系统节点出错检测和失效恢复。

4.2 MapReduce 的工作原理

MapReduce 作业的执行涉及 4 个独立的实体。

(1) Jobclient：编写 MapReduce 客户端程序、配置作业和提交作业，是开发人员需要完成的工作。

(2) JobTracker：初始化作业、分配作业，与 TaskTracker 通信，协调整个作业的执行。

(3) TaskTracker：保持与 JobTracker 的通信，在分配的数据片段上执行 Map 或 Reduce 任务。TaskTracker 和 JobTracker 的不同是在执行任务时 TaskTracker 可以有多个，JobTracker 则只会有一个。

(4) HDFS：保存作业的数据、配置信息等，最后的结果也保存在 HDFS 上面。

MapReduce 工作原理图如图 4-3 所示。

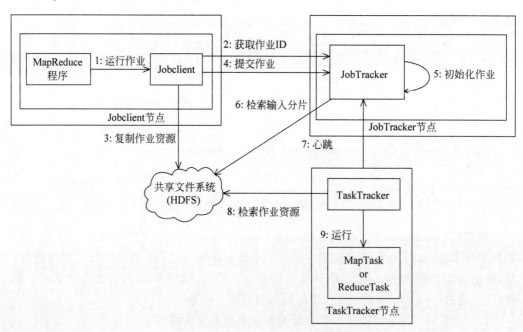

图 4-3 MapReduce 工作原理图

下面从 Jobclient、JobTracker 和 TaskTracker 的角度来分析 MapReduce 的工作原理如下：

(1) MapReduce 程序。客户端(Jobclient)要编写好 MapReduce 程序，配置 MapReduce 的 Job 作业。

（2）启动 Job。启动 Job 是告知 JobTracker 上要运行作业，这个时候 JobTracker 就会返回给客户端一个新的 JobID 值，接下来它会做检查操作，这个检查就是确定输出目录是否存在，如果存在那么 Job 就不能正常运行下去，JobTracker 会抛出错误给客户端，接下来还要检查输入目录是否存在，如果不存在同样抛出错误，如果存在 JobTracker 会根据输入计算输入分片（Input Split），如果分片计算不出来也会抛出错误。之后 JobTracker 就会配置 Job 需要的资源了。拿到 JobID 后，将运行作业所需的资源文件复制到 HDFS，包括 MapReduce 程序打包的 JAR 文件、配置文件和计算所得的输入分片信息。这些文件都存放在 JobTracker 为该作业创建的文件夹中，文件夹名为该作业的 JobID。JAR 文件默认会有 10 个副本（Mapred.submit.replication 属性控制）；输入分片信息决定 JobTracker 应该为这个作业启动多少个 Map 任务等信息。当资源文件夹创建完毕后，客户端会提交 Job 告知 JobTracker 已将所需资源写入 HDFS。

（3）执行 Job。分配好资源后，JobTracker 接收提交 Job 请求后就会初始化作业，初始化主要是将 Job 放入一个内部的队列，等待作业调度器对其进行调度。当作业调度器根据自己的调度算法调度到该作业时，作业调度器会创建一个正在运行的 Job 对象，以便 JobTracker 跟踪 Job 的状态和进程。创建 Job 对象时作业调度器会获取 HDFS 文件夹中的输入分片信息，根据分片信息为每个 Input Split 创建一个 Map 任务，并将 Map 任务分配给 TaskTracker 执行。对于 Map 和 Reduce 任务，TaskTracker 根据主机核的数量和内存的大小确定一定数量的 Map 槽和 Reduce 槽。

（4）执行任务。在任务执行时 JobTracker 可以通过心跳机制监控 TaskTracker 的状态和进度，同时也能计算出整个 Job 的状态和进度，而 TaskTracker 也可以本地监控自己的状态和进度。TaskTracker 每隔一段时间会给 JobTracker 发送一个心跳，告诉 JobTracker 它依然在运行，同时心跳中还携带很多信息，比如当前 Map 任务完成的进度等。当 JobTracker 获得了最后一个完成指定任务的 TaskTracker 操作成功的通知时，JobTracker 会把整个 Job 状态置为成功，然后当客户端查询 Job 运行状态时（异步操作），客户端会查到 Job 完成的通知。如果 Job 中途失败，MapReduce 也会有相应机制处理。

4.3　MapReduce 的运行机制

MapReduce 的运行机制如图 4-4 所示。

首先文件要存储在 HDFS 中，系统把每个文件切分成多个一定大小的 Block（默认 3 个备份）存储在多个节点（DataNode）上。

（1）输入和拆分。输入和拆分不属于 Map 和 Reduce 的主要过程，但属于整个计算框架消耗时间的一部分，该部分会为正式的 Map 过程准备数据。

① 分片（Split）操作。MapReduce 框架使用 InputFormat 基础类做 Map 前的预处理，比如验证输入的格式是否符合输入定义；然后，将输入文件切分为逻辑上的多个输入分片，输入分片是 MapReduce 对文件进行处理和运算的输入单位，只是一个逻辑概念。在进行 Map 计算之前，MapReduce 会根据输入文件计算输入分片，每个输入分片针对一个 Map 任务，输入分片存储的并非数据本身，而是一个分片长度和一个记录数据的位置

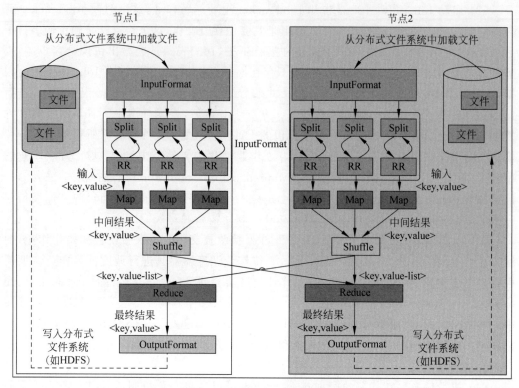

图 4-4　MapReduce 的运行机制

的数组。分片操作将源文件的内容分片形成一系列的输入分片,每个输入分片中存储着该分片的数据信息(例如,文件块信息、起始位置、数据长度、所在节点列表等),并不是对文件实际分割成多个小文件,每个输入分片都由一个 Map 任务进行后续处理。

② 输入分片。输入分片往往和 HDFS 的 Block(块)关系很密切,假如设定的块的大小是 64MB,如果输入有 3 个文件,大小分别是 3MB、65MB 和 127MB,那么 MapReduce 会把 3MB 文件分为一个输入分片,65MB 则是两个输入分片,而 127MB 也是两个输入分片,那么就会有 5 个 Map 任务将执行,而且每个 Map 执行的数据大小不均,这个也是 MapReduce 优化计算的一个关键点。

③ 数据格式化(Format)操作。将划分好的输入分片格式化成键值对形式的数据。其中 key 为偏移量,value 为每一行的内容。值得注意的是,在 Map 任务执行过程中,会不停地执行数据格式化操作,每生成一个键值对就会将其传入 Map 进行处理。所以 Map 和数据格式化操作并不存在前后时间差,而是同时进行的。这里具体涉及 RecordReader 类,其作用是从分片中每读取一条记录,就调用一次 Map()函数。因为输入分片是逻辑切分而非物理切分,所以还需通过 RecordReader 根据输入分片中的信息来处理输入分片中的具体记录,加载数据并转换为适合 Map 任务读取的键值对,输入给 Map 任务。例如记录"we are studying at school"作为参数 v,调用 Map(v),然后继续这个过程,读取下一条记录直到输入分片尾部。

(2) Map 阶段。Map 阶段是开发人员编写好 Map()函数,因此 Map()函数效率相对好控制,而且一般 Map 操作都是本地化操作,也就是在数据存储节点上进行。在 HDFS 中,文件数据是被复制多份的,所以计算将会选择拥有此数据的最空闲的节点。比如记录"we are studying at school",调用执行一次 Map(),在内存中增加数据:{"we":1},{"are":1},{"studying":1},{"at":1},{"school":1}。

（3）Shuffle 阶段。Shuffle 阶段是指 Map 产生的直接输出结果,经过一系列的处理,成为最终的 Reduce 直接输入的数据为止的整个过程。这是 MapReduce 的核心过程。该过程可以分为两个阶段。

① Map 端的 Shuffle。由 Map 处理后的结果并不会直接写入磁盘中,而是会在内存中开启一个环形内存缓冲区,先将 Map 的处理结果写入缓冲区中,这个缓冲区默认大小是 100MB,并且在配置文件中为这个缓冲区设定了一个阈值,默认是 0.80（这个大小和阈值都是可以在配置文件中进行配置的）,同时 Map 还会为输出操作启动一个守护线程,如果缓冲区的内存达到了阈值的 80%,这个守护线程就会把内容写到磁盘上,这个过程叫溢写（Spill）,另外的 20% 内存可以继续读入要写进磁盘的数据,写入磁盘和读入内存操作是互不干扰的,如果缓存区满了,那么 Map 就会阻塞写入内存的操作,让写入磁盘操作完成后再继续执行读入内存操作。每次溢写操作也就是写入磁盘操作时就会写一个溢出文件,也就是说 Map 输出有几次,溢写就会产生多少个溢出文件。随着数据不断读入内存缓冲区,会溢出并产生多个小文件,等 Map 输出全部完成后,Map 会将这些输出文件合并为一个大文件。同时,在数据溢出转储到磁盘这一过程是复杂的,并不是直接写入磁盘,而是在写入磁盘前 Map 会对数据执行分区（Partition）、排序（Sort）、合并（Combiner）和归并（Merge）等操作。Map 端的 Shuffle 示意图如图 4-5 所示。

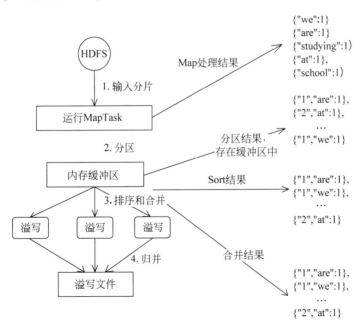

图 4-5　Map 端的 Shuffle 示意图

a. 分区。在数据写入内存时,决定数据由哪个 Reduce 处理,从而分区。比如采用 Hash 法,有 n 个 Reducer,那么数据{"are",1}的 key "are"对 n 进行取模,返回 m,从而生成{partition,key,value}。其实 Partitioner 操作和 Map 阶段的输入分片很像,一个 Partitioner 对应一个 Reduce 作业,如果 MapReduce 操作只有一个 Reduce 操作,那么 Partitioner 就只有一个,如果有多个 Reduce 操作,那么 Partitioner 对应的就会有多个。因此 Partitioner 就是 Reduce 的输入分片。每个 Map 的处理结果和 Partition 处理的 <key,value> 键值对结果都保存在缓存 MemoryBuffer 中。缓冲区中的数据格式为 partition key value 三元组数据,如{"1","are":1},{"2","at":1},…,{"1","we":1}。

b. 排序。在溢出的数据写入磁盘前会对数据按照 key 进行排序操作,默认算法为快速排序,第一关键字为分区号,第二关键字为 key。这个是在写入磁盘操作时候进行,不是在写入内存时进行的。例如,缓冲区数据{"1","are":1},{"2","at":1},…,{"1","we":1},排序后为{"1","are":1}{"1","we":1},…,{"2","at":1}。

c. 合并。合并阶段是开发人员可以选择的。数据合并在 Reduce 计算前,对相同的 key 的数据,value 值合并,减少输出传输量,Combiner() 函数事实上是本地化的 Reducer() 函数。但是合并操作是有风险的,使用它的原则是合并的输入不会影响到 Reduce 计算的最终输入。如果计算只是求总数、最大值、最小值,则可以使用合并,但是做平均值或求中值计算使用合并,最终的 Reduce 计算结果就会出错。例如,两个 Mapper 端使用合并平均值操作,如下所示:

Mapper1:3 5 7 —>(3+5+7)/3=5

Mapper2:2 6 —>(2+6)/2=4

Reducer:(5+4)/2=9/2,不等于(3+5+7+2+6)/5=23/5

d. 归并。每次溢写会生成一个溢写文件,这些溢写文件最终需要被归并成一个大文件。归并的意思是生成 key 和对应的 value-list。在 Map 任务全部结束之前进行归并,归并得到一个大的文件,放在本地磁盘。

合并和归并是有区别的,例如,两个键值对<"a",1>和<"a",1>,如果合并,会得到<"a",2>;如果归并,会得到<"a",<1,1>>。

② Reduce 阶段的 Shuffer。由于 Map 和 Reduce 往往不在同一个节点上运行,因此 Reduce 需要从多个节点上下载 Map 的结果数据,多个节点的 Map 中相同分区内的数据被复制到同一个 Reduce 上,并对这些数据进行处理,然后才能作为 Reduce 的输入数据被 Reduce 处理。Reduce 阶段的 Shuffer 示意图如图 4-6 所示。

a. 领取数据。Reduce 端可能从 n 个 Map 的结果中获取数据,而这些 Map 的执行速度不尽相同,当其中一个 Map 运行结束时,Reduce 就会从 JobTracker 中获取该信息。Map 运行结束后 TaskTracker 会得到消息,进而将消息汇报给 JobTracker,Reduce 定时从 JobTracker 获取该信息,Reduce 端默认有 5 个数据复制线程从 Map 端复制数据。Reduce 任务通过 RPC 向 JobTracker 询问 Map 任务是否已经完成,若完成,则复制数据。

b. 归并数据。Reduce 复制数据先放入缓存,内存缓冲区满时,也通过分区和合并,将数据溢写到磁盘文件中。如果形成了多个磁盘文件还会进行归并,最后一次归并的结果作为 Reduce 的输入而不是写入磁盘中。文件中的键值对是排序的。当数据很少时,

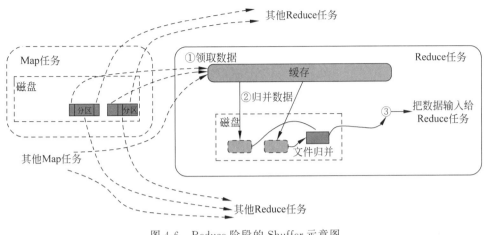

图 4-6　Reduce 阶段的 Shuffer 示意图

不需要溢写到磁盘，直接在缓存中归并。

c. 把数据输入给 Reduce 任务。Reduce()函数和 Map()函数一样也是开发人员编写的，最终结果是存储在 HDFS 上的。每个 Reduce 进程会对应一个输出文件，名称以 part-开头。

4.4　MapReduce 数据本地化

HDFS 和 MapReduce 是 Hadoop 的核心设计。HDFS 是存储的基础，在数据层面上提供了海量数据存储的支持。而 MapReduce 是在数据的上一层，通过编写 MapReduce 程序对海量数据进行计算处理。

在 HDFS 中，NameNode 是文件系统的名字节点进程，DataNode 是文件系统的数据节点进程。MapReduce 计算框架中负责计算任务调度的 JobTracker 对应 HDFS 的 NameNode 角色，只不过一个负责计算任务调度，一个负责存储任务调度。MapReduce 中负责真正计算任务的 TaskTracker 对应 HDFS 的 DataNode 角色，一个负责计算，一个负责管理存储数据。

考虑到"本地化原则"，一般地，将 NameNode 和 JobTracker 部署到同一台机器上，各个 DataNode 和 TaskTracker 也同样部署到同一台机器上。

这样做的目的是将 Map 任务分配给含有该 Map 处理的数据块的 TaskTracker 上，也就是在输入分片所对应的数据块所在的存储节点上，由该节点的 TaskTracker 执行 Map 任务，同时将程序 JAR 包复制到该 TaskTracker 上来运行，这叫作"计算移动，数据不移动"。而分配 Reduce 任务时并不考虑数据本地化。

4.5　MapReduce 编程

4.5.1　MapReduce 运行模式

MapReduce 运行模式主要有以下两种。

（1）本地运行模式：在当前的开发环境模拟 MapReduce 执行环境，处理的数据及输出结果在本地操作系统。

（2）集群运行模式：把 MapReduce 程序打成一个 JAR 包，提交至 YARN 集群上运行任务。由于 YARN 集群负责资源管理和任务调度，程序会被框架分发到集群中的节点上并发执行，因此处理的数据和输出结果都在 HDFS 中。

4.5.2 MapReduce 编程组件与数据类型

MapReduce 编程组件主要有以下 6 种。

（1）InputFormat 组件。其主要用于描述输入数据的格式，它提供两个功能，分别是数据切分和为 Mapper 提供输入数据。

（2）OutputFormat 组件。OutputFormat 组件是一个描述 MapReduce 程序输出格式和规范的抽象类。

（3）Combiner 组件。Combiner 组件的作用就是对 Map 阶段的输出的重复数据先做一次合并计算，然后把新的<key,value>作为 Reduce 阶段的输入。

（4）Mapper 组件。Hadoop 提供的 Mapper 类是实现 Map 任务的一个抽象基类，该基类提供了一个 Map()方法。

（5）Reducer 组件。Map 过程输出的键值对将由 Reducer 组件进行合并处理，最终以某种形式的结果输出。

（6）Partitioner 组件。Partitioner 组件可以让 Map 对 key 进行分区，从而可以根据不同的 key 分发到不同的 Reduce 中去处理，其目的就是将 key 均匀分布在 ReduceTask 上。

Hadoop 提供 Java 编程的数据类型，如表 4-1 所示。这些数据类型都实现了 WritableComparable 接口，以便使用这些类型定义的数据可以被序列化进行网络传输和文件存储，以及进行大小比较。

表 4-1 Hadoop 提供 Java 编程数据类型

数 据 类 型	描 述
BooleanWritable	标准布尔型数
ByteWritable	单字节数
DoubleWritable	双字节数
FloatWritable	浮点数
IntWritable	整型数
LongWritable	长整型数
Text	使用 UTF-8 格式存储的文本
NullWritable	当 key 或 value 为空时使用

4.6 MapReduce 编程示例

4.6.1 单词计数

该程序的完整代码可以在 Hadoop 安装包的"src/examples"目录下找到。单词计数主要完成功能是统计一系列文本文件中每个单词出现的次数。

(1) 程序解析。

① 首先 MapReduce 将文件拆分成 Splits，由于测试用的文件较小，只有两行文字，因此每个文件为一个 Split，并将文件按行分割形成<key,value>对，MapReduce 文件拆分示意图如图 4-7 所示。这一步由 MapReduce 框架自动完成，其中偏移量(即 key 值)包括了回车(即换行符)所占的字符数(Windows 和 Linux 环境不同)。

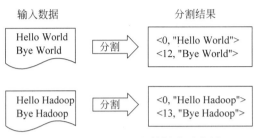

图 4-7　MapReduce 文件拆分示意图

② 将分割好的<key,value>对交给用户定义的 Map()方法进行处理，生成新的<key,value>对。Map()方法处理过程示意图如图 4-8 所示。

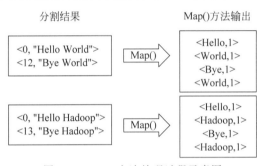

图 4-8　Map()方法处理过程示意图

③ 得到 Map()方法输出的<key,value>对后，Mapper 会将它们按照 key 值进行排序，并执行合并过程，将 key 至相同 value 值累加，得到 Mapper 的最终输出结果。

④ Reducer 先对从 Mapper 接收的数据进行排序，再交由用户自定义的 Reduce()方法进行处理，得到新的<key,value>对，并作为 WordCount 的输出结果。Reduce()方法处理过程示意图，如图 4-9 所示。

图 4-9　Reduce()方法处理过程示意图

(2) 程序编写过程。

① 进行 IP 和主机名的映射。例如，在 Windows 操作系统的 Hosts 文件中加入 Hadoop 集群的 IP 和主机名的映射，此处三个主机为第 2 章安装 Hadoop 时配置的 IP 和主机名。

```
172.16.106.69    bigdata01
172.16.106.70    bigdata02
172.16.106.71    bigdata03
```

② 新建工程。使用 IntelliJ IDEA 新建一个 MAVEN 工程，工程名称为 WordCout，新建 MAVEN 工程界面如图 4-10 所示。

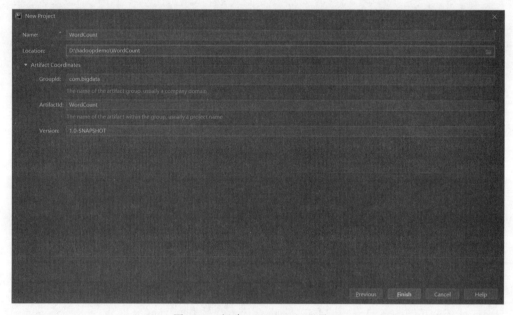

图 4-10　新建 MAVEN 工程界面

③ 导入依赖包，pom.xml 文件配置内容如下：

```xml
<?xml version = "1.0" encoding = "UTF-8"?>
<project xmlns = "http://maven.apache.org/POM/4.0.0"
    xmlns:xsi = "http://www.w3.org/2001/XMLSchema-instance"
    xsi:schemaLocation = "http://maven.apache.org/POM/4.0.0
    http://maven.apache.org/xsd/maven-4.0.0.xsd">
    <modelVersion>4.0.0</modelVersion>
    <groupId>com.bigdata</groupId>
    <artifactId>WordCount</artifactId>
    <version>1.0-SNAPSHOT</version>
    <properties>
        <project.build.sourceEncoding>UTF-8</project.build.sourceEncoding>
        <hadoop.version>2.7.7</hadoop.version>
    </properties>
    <dependencies>
```

```xml
<dependency>
    <groupId>org.apache.hadoop</groupId>
    <artifactId>hadoop-common</artifactId>
    <version>${hadoop.version}</version>
</dependency>
<dependency>
    <groupId>org.apache.hadoop</groupId>
    <artifactId>hadoop-hdfs</artifactId>
    <version>${hadoop.version}</version>
</dependency>
<dependency>
    <groupId>org.apache.hadoop</groupId>
    <artifactId>hadoop-MapReduce-client-core</artifactId>
    <version>${hadoop.version}</version>
</dependency>
<dependency>
    <groupId>org.apache.hadoop</groupId>
    <artifactId>hadoop-client</artifactId>
    <version>${hadoop.version}</version>
</dependency>
</dependencies>
<build>
<plugins>
    <plugin>
        <groupId>org.apache.maven.plugins</groupId>
        <artifactId>maven-assembly-plugin</artifactId>
        <version>2.5.5</version>
        <configuration>
            <archive>
                <manifest>
                    <addClasspath>true</addClasspath>
                    <classpathPrefix>lib/</classpathPrefix>
                    <mainClass>com.bigdata.wc.WordCount</mainClass>
                </manifest>
            </archive>
            <descriptorRefs>
                <descriptorRef>jar-with-dependencies</descriptorRef>
            </descriptorRefs>
        </configuration>
        <executions>
            <execution>
                <id>make-assembly</id>
                <phase>package</phase>
                <goals>
                    <goal>single</goal>
                </goals>
            </execution>
        </executions>
    </plugin>
```

```
    </plugins>
  </build>
</project>
```

④ 编写 WordCount 程序,代码及说明如下:

```java
import java.io.IOException;
import java.util.StringTokenizer;
import org.apache.hadoop.conf.Configuration;
import org.apache.hadoop.fs.Path;
import org.apache.hadoop.io.IntWritable;
import org.apache.hadoop.io.Text;
import org.apache.hadoop.MapReduce.Job;
import org.apache.hadoop.MapReduce.Mapper;
import org.apache.hadoop.MapReduce.Reducer;
import org.apache.hadoop.MapReduce.lib.input.FileInputFormat;
import org.apache.hadoop.MapReduce.lib.output.FileOutputFormat;
import org.apache.hadoop.util.GenericOptionsParser;
public class WordCount{
    /**
     * 建立 Mapper 类 TokenizerMapper,继承自泛型类 Mapper
     * Mapper 类:实现了 Map 功能基类
     * Mapper 接口:
     * WritableComparable 接口: 实现 WritableComparable 的类可以相互比较。所有被用作
       key 的类应该实现此接口。
     * Reporter 则可用于报告整个应用的运行进度,本例中未使用。
     */
    public static class TokenizerMapper
            extends Mapper<Object, Text, Text, IntWritable> {
        private final static IntWritable one = new IntWritable(1);
        private Text word = new Text();
    /**
     * Mapper 中的 Map()方法:
     * void Map(K1 key,V1 value, Context context)
     * 映射一个单个的输入<key,value>对到一个中间的<key,value>对
     * 输出对不需要和输入对是相同的类型,输入对可以映射到 0 个或多个输出对。
     * Context: 收集 Mapper 输出的<key,value>对。
     * Context 的 write(k,v)方法:增加一个(key,value)对到 context
     * 开发人员主要编写 Map()和 Reduce()函数. 这个 Map()函数使用 StringTokenizer()
       函数对字符串进行分割,通过 write()方法把单词存入 word 中
     * write()方法存入(单词,1)这样的二元组到 context 中
     */
        public void Map(Object key, Text value, Context context) throws IOException,
InterruptedException {
            StringTokenizer itr = new StringTokenizer(value.toString());
            while (itr.hasMoreTokens()){
                word.set(itr.nextToken());
                System.out.println(word);
                context.write(word, one);
            }
```

```java
        }
    }
    /**
     * IntSumReducer 类中的 Reduce()方法:
     * void Reduce(Text key,Iterable<IntWritable> values,Context context)
     * <key,value>来自 Map()函数中的 context,可能经过了进一步处理(combiner),同样通
       过 context 输出
     */
    public static class IntSumReducer extends Reducer<Text,IntWritable,Text,IntWritable> {
    private IntWritable result = new IntWritable();
    public void Reduce(Text key, Iterable<IntWritable> values,Context context) throws
     IOException,InterruptedException {
            int sum = 0;
            for (IntWritable val : values){
                sum += val.get();
            }
            result.set(sum);
            context.write(key, result);
        }
    }
    public static void main(String[] args) throws Exception{
    System.setProperty("hadoop.home.dir", "D:\\hadoop-2.7.0");
    System.setProperty("HADOOP_USER_NAME","root");
    /**
     * Configuration: Map/Reduce 的配置类,描述 Hadoop 框架 Map-Reduce 执行工作
     */
    //新建配置类
    Configuration conf = new Configuration();
    //配置 resourcemanager 地址
    conf.set("yarn.resourcemanager.address",bigdata01:8032");
    //允许 DataNode 以主机名访问
    conf.set("dfs.client.use.datanode.hostname","true");
    //配置 HDFS 访问地址
    conf.set("fs.defaultFS", "hdfs://bigdata02:9000/");
    //配置 MapReduce 提交方式为跨平台提交
    conf.set("MapReduce.app-submission.cross-platform", "true");
    //设置 Job 提交 YARN 去运行
    conf.set("MapReduce.framework.name","yarn");
    //设置 JAR 本地路径
    conf.set("Mapred.jar","D:\hadoopdemo\WordCount\target\WordCount-1.0-SNAPSHOT-jar-with-dependencies.jar");
    //取得输入参数值
    String[] otherArgs = new GenericOptionsParser(conf, args).getRemainingArgs();
    if (otherArgs.length < 2){
            System.err.println("Usage: wordcount <in> [<in>...] <out>");
            System.exit(2);
    }
    //设置一个用户定义的 Job 名称
    Job job = Job.getInstance(conf, "word count");
```

```
//为 Job 设置类名
job.setJarByClass(WordCount.class);
//为 Job 设置 Mapper 类
job.setMapperClass(TokenizerMapper.class);
//为 Job 设置 Combiner 类
job.setCombinerClass(IntSumReducer.class);
//为 Job 设置 Reducer 类
job.setReducerClass(IntSumReducer.class);
//为 Job 的输出数据设置 Key 类
job.setOutputKeyClass(Text.class);
//为 Job 输出设置 Value 类
job.setOutputValueClass(IntWritable.class);
for (int i = 0; i < otherArgs.length - 1; ++i){
    //为 Job 设置输入路径
    FileInputFormat.setInputPaths(job,new Path(otherArgs[i]));
}
//为 Job 设置输出路径
FileOutputFormat.setOutputPath(job,new Path(otherArgs[otherArgs.length - 1]));
//运行 Job
System.exit(job.waitForCompletion(true) ? 0 : 1);
    }
}
```

⑤ 打 JAR 包。WordCount 打 JAR 包界面如图 4-11 所示。

图 4-11 WordCount 打 JAR 包界面

⑥ 配置 WordCount 运行参数。WordCount 运行需要两个参数：第一个参数是输入文件 hdfs://bigdata02:9000/input/；第二个参数是结果输出路径 hdfs://bigdata02:9000/output。WordCount 运行参数配置界面如图 4-12 所示。

图 4-12　WordCount 运行参数配置界面

⑦ 运行 WordCount 程序。运行成功后，WordCount 控制台输出结果如图 4-13 所示。其表明 MapReduce 任务成功提交到远程集群运行。

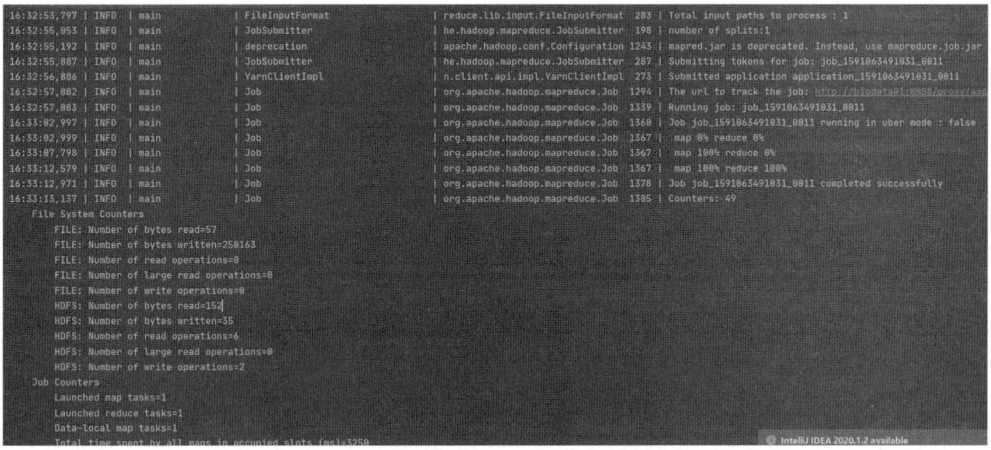

图 4-13　WordCount 控制台输出结果

⑧ 验证结果。通过浏览器访问 http://172.6.106.69:50070，在 NameNode 监控界面，选择 utillities→browse the file system 选项，可以看到刚才提交 WordCount 程序在 Hadoop 集群生成的 output 目录。WordCount 程序生成的目录如图 4-14 所示。

打开这个目录，可以看到一系列文件，WordCount 程序运行结果如图 4-15 所示。其中 part-r-00000 中存入的是程序的运行结果。

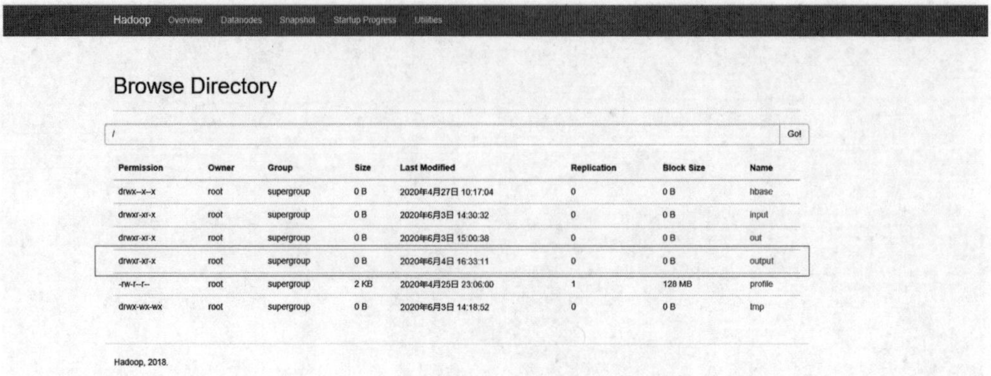

图 4-14 WordCount 程序生成的目录

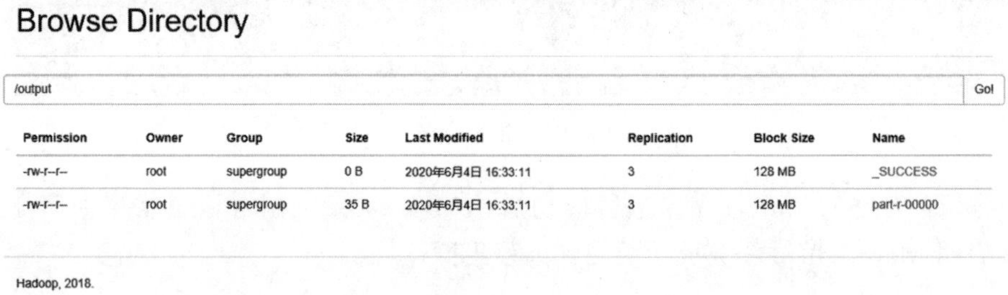

图 4-15 WordCount 程序运行结果

4.6.2 倒排索引

倒排索引(Inverted Index)是文档检索系统中最常用的数据结构,被广泛应用于全文搜索引擎。倒排索引主要用来存储某个单词(或词组)在一组文档中的存储位置的映射,提供了可以根据内容来查找文档的方式,而不是根据文档来确定内容,因此称为倒排索引。带有倒排索引的文件称为倒排索引文件,简称倒排文件(Inverted File)。

(1) 程序解析。

现有 file1.txt、file2.txt 和 file3.txt 三个源文件,需要使用倒排索引的方式对这三个源文件内容实现倒排索引,并将最后的倒排索引文件输出。倒排索引功能实现示意图如图 4-16 所示。

首先,使用默认的 TextInputFormat 类对每个输入文件进行处理,得到文本中每行的偏移量及其内容。Map 过程首先分析输入的 <key,value> 键值对,经过处理可以得到倒排索引中需要的三个信息:单词、文档名称和词频。倒排索引 Map 过程示意图如图 4-17 所示。

经过 Map 阶段数据转换后,同一个文档中相同的单词会出现多个的情况,而单纯依

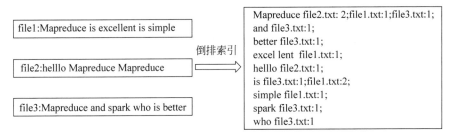

图 4-16　倒排索引功能实现示意图

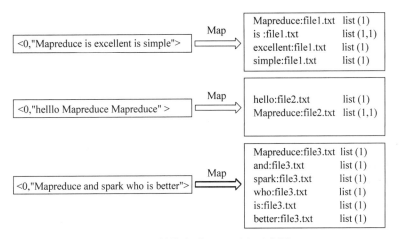

图 4-17　倒排索引 Map 过程示意图

靠后续 Reduce 阶段无法同时完成词频统计和生成文档列表，所以必须增加一个 Combine 过程，先完成每一个文档的词频统计。倒排索引 Combine 过程示意图如图 4-18 所示。

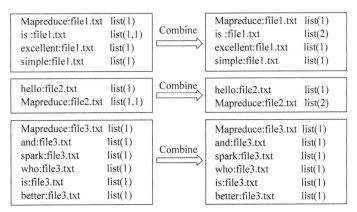

图 4-18　倒排索引 Combine 过程示意图

经过上述两个阶段的处理后，Reduce 阶段只需将所有文件中相同 key 值的 value 值进行统计，并组合成倒排索引文件所需的格式。倒排索引 Reduce 过程示意图如图 4-19 所示。

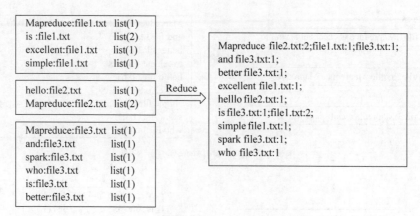

图 4-19 倒排索引 Reduce 过程示意图

(2) 程序编写过程。

① 数据准备。在 Hadoop 集群中建立目录/InvertedIndex/input,并将 file1. txt、file2. txt、file3. txt 文件编辑好后上传到 Hadoop。Hadoop Shell 命令如下:

```
hadoop fs - mkdir /InvertedIndex/
hadoop fs - mkdir /InvertedIndex/input
hadoop fs - put /file1.txt /InvertedIndex/input/
hadoop fs - put /file2.txt /InvertedIndex/input/
hadoop fs - put /file3.txt /InvertedIndex/input/
```

② 建立 MAVEN 工程和导入依赖包。除了要修改 pom. xml 中 mainClass 路径外,其他与 WordCount 相同。

③ 编写 Map 阶段代码,新建 Java 类 InvertedIndexMapper,相关代码及说明如下:

```
package com.mr.InvertedIndex;
import java.io.IOException;
import org.apache.commons.lang.StringUtils;
import org.apache.hadoop.io.LongWritable;
import org.apache.hadoop.io.Text;
import org.apache.hadoop.MapReduce.Mapper;
import org.apache.hadoop.MapReduce.lib.input.FileSplit;
public class InvertedIndexMapper extends Mapper<LongWritable,Text,Text,Text> {
    //存储单词和文件名组合
    private static Text keyInfo = new Text();
    //存储词频,初始化为 1
    private static final Text valueInfo = new Text("1");
    @Override
    protected void Map(LongWritable key, Text value, Context context) throws IOException, InterruptedException {
        String line = value.toString();
        //得到字段数组
        String[] fields = StringUtils.split(line, " ");
        //得到这行数据所在的文件切片
        FileSplit fileSplit = (FileSplit) context.getInputSplit();
```

```
        //根据文件切片得到文件名
        String fileName = fileSplit.getPath().getName();
        for (String field : fields){
            //key 值由单词和 URL 组成,如"MapReduce:file1"
            keyInfo.set(field + ":" + fileName);
            context.write(keyInfo, valueInfo);
        }
    }
}
```

④ 编写 Combine 阶段代码,新建 Java 类 InvertedIndexCombiner,相关代码及说明如下:

```
import java.io.IOException;
import org.apache.hadoop.io.Text;
import org.apache.hadoop.mapreduce.Reducer;
public class InvertedIndexCombiner extends Reducer<Text, Text, Text, Text> {
    private static Text info = new Text();
    //输入:<MapReduce:file3 {1,1,…}>
    //输出:<MapReduce file3:2>
    @Override
    protected void reduce(Text key, Iterable<Text> values, Context context) throws IOException,
InterruptedException {
        //统计词频
        int sum = 0;
        for (Text value : values) {
            sum += Integer.parseInt(value.toString());
        }
        //查找分割位置
        int splitIndex = key.toString().indexOf(":");
        //重新设置 value 值由 URL 和词频组成
        info.set(key.toString().substring(splitIndex + 1) + ":" + sum);
        //重新设置 key 值为单词
        key.set(key.toString().substring(0, splitIndex));
        context.write(key, info);
    }
}
```

⑤ 编写 Reduce 阶段代码,新建 Java 类 InvertedIndexReducer,相关代码及说明如下:

```
package com.mr.InvertedIndex;
import java.io.IOException;
import org.apache.hadoop.io.Text;
import org.apache.hadoop.MapReduce.Reducer;
public class InvertedIndexReducer extends Reducer<Text,Text,Text,Text> {
        private static Text result = new Text();
    //输入<MapReduce file3:2>
    //输出<MapReduce file1:1;file2:1;file3:2;>
    @Override
```

```java
protected void Reduce(Text key, Iterable<Text> values, Context context)
        throws IOException,InterruptedException {
    //生成文档列表
    String fileList = new String();
    for (Text value : values) {
        fileList += value.toString() + ";";
    }
    result.set(fileList);
    context.write(key,result);
}
```

⑥ 编写 Job 配置运行代码,新建 Java 类 InvertedIndexRunner,相关代码及说明如下:

```java
package com.mr.InvertedIndex;
import java.io.IOException;
import org.apache.hadoop.conf.Configuration;
import org.apache.hadoop.fs.Path;
import org.apache.hadoop.io.Text;
import org.apache.hadoop.MapReduce.Job;
import org.apache.hadoop.MapReduce.lib.input.FileInputFormat;
import org.apache.hadoop.MapReduce.lib.output.FileOutputFormat;
public class InvertedIndexRunner{
    public static void main(String[] args) throws IOException,
        ClassNotFoundException,InterruptedException {
        //配置本地 Hadoop 路径
        System.setProperty("hadoop.home.dir", "D:\hadoop-2.7.0");
        //配置本地 Hadoop 用户名
        System.setProperty("HADOOP_USER_NAME","root");
        //新建配置类
        Configuration conf = new Configuration();
        //配置 RM 地址
        conf.set("yarn.resourcemanager.address","bigdata01:8032");
        //允许 DataNode 以主机名访问
        conf.set("dfs.client.use.datanode.hostname", "true");
        //配置 HDFS 访问地址
        conf.set("fs.defaultFS","hdfs://bigdata02:9000/");
        //配置 MapReduce 提交方式为跨平台提交
        conf.set("MapReduce.app-submission.cross-platform", "true");
        //设置 Job 提交 YARN 去运行
        conf.set("MapReduce.framework.name", "yarn");
        //配置 resourcemanager 主机名
        conf.set("yarn.resourcemanager.hostname","bigdata01");
        //设置 JAR 本地路径
        conf.set("Mapred.jar", "D:\hadoopdemo\mrdemo\target\mrdemo-1.0-SNAPSHOT-jar-with-dependencies.jar");
        //构建 Job 实例
        Job job = Job.getInstance(conf);
        //为 Job 设置类名
        job.setJarByClass(InvertedIndexRunner.class);
        //为 Job 设置 Mapper 类
        job.setMapperClass(InvertedIndexMapper.class);
```

```java
        //为 Job 设置 Combiner 类
        job.setCombinerClass(InvertedIndexCombiner.class);
        //为 Job 设置 Reducer 类
        job.setReducerClass(InvertedIndexReducer.class);
        //为 Job 的输出数据设置 Key 类
        job.setOutputKeyClass(Text.class);
        //为 Job 输出设置 Value 类
        job.setOutputValueClass(Text.class);
        //为 Job 设置输入路径
        FileInputFormat.setInputPaths(job,new
            Path("hdfs://bigdata02:9000/InvertedIndex/input"));
        //指定处理完成之后的结果所保存的位置
        FileOutputFormat.setOutputPath(job,new
            Path("hdfs://bigdata02:9000/InvertedIndex/output"));
        //向 YARN 集群提交这个 Job
        boolean res = job.waitForCompletion(true);
        System.exit(res ? 0 : 1);
    }
}
```

⑦ 打 JAR 包，与 WordCoumt 步骤相同。

⑧ 运行 InvertedIndexRunner。运行 InvertedIndexRunner 控制台输出结果如图 4-20 所示。

图 4-20　运行 InvertedIndexRunner 控制台输出结果

⑨ 验证结果。查看 Hadoop，可以看到程序在 Hadoop 集群上生成的 output 目录，其中 part-r-00000 里存入的是程序的运行结果。倒排索引程序运行结果如图 4-21 所示。

```
[root@bigdata01 /]# hadoop fs -ls /InvertedIndex/output/
Found 2 items
-rw-r--r--   3 root supergroup          0 2020-06-05 12:48 /InvertedIndex/output/_SUCCESS
-rw-r--r--   3 root supergroup        288 2020-06-05 12:48 /InvertedIndex/output/part-r-00000
```

图 4-21　倒排索引程序运行结果

4.7　本章小结

本章主要介绍了 Hadoop 的分布式计算框架 MapReduce，分析了 MapReduce 的核心思想、工作原理和运行机制。重点掌握 MapReduce 的核心过程 Shuffle 阶段，最后通过详细分析和编写单词计数代码和倒排索引代码，让读者掌握 MapReduce 的原理和计算过程。

第 5 章

Hive数据仓库

5.1 Hive 概述

5.1.1 Hive 简介

Hive 起源于 Facebook 公司,Facebook 公司有大量的日志数据,而 Hadoop 是实现了 MapReduce 模式开源的分布式并行计算的框架,可轻松处理大规模数据。而对于 Java 语言熟悉的工程师来说开发 MapReduce 程序很容易,但对于其他语言使用者则难度较大。因此,Facebook 开发团队想设计一种使用 SQL 对日志数据查询分析的工具,Hive 就诞生于此。只要懂 SQL,就能够利用 Hive 胜任大数据分析方面的工作,节省了开发人员的学习成本。

Hive 是建立在 Hadoop 文件系统上的数据仓库分析系统,它提供了一系列工具,能够对存储在 HDFS 中的数据进行数据提取、转换和加载,可以存储、查询和分析存储在 Hadoop 中的大规模数据。Hive 定义简单的类 SQL(即 HQL),可以将结构化的数据文件映射为一张数据表,允许熟悉 SQL 的用户查询数据。

Hive 采用了 SQL 的查询语言 HQL,因此很容易将 Hive 理解为数据库。从结构上来看,Hive 和数据库除了拥有类似的查询语言外,再无类似之处,Hive 与 MySQL 对比如表 5-1 所示。

表 5-1 Hive 与 MySQL 对比

对 比 项	Hive	MySQL
查询语言	HQL	SQL
数据存储位置	HDFS	块设备、本地文件系统
数据格式	用户定义	系统决定

续表

对 比 项	Hive	MySQL
数据更新	不支持	支持
事务	不支持	支持
执行延迟	高	低
可扩展性	高	低
数据规模	大	小

5.1.2 Hive 的架构

Hive 的架构如图 5-1 所示。

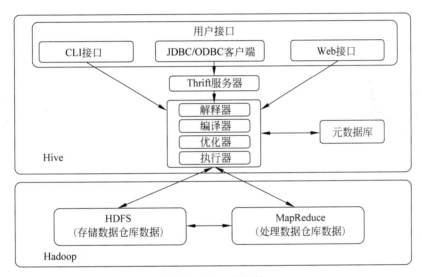

图 5-1 Hive 的架构

(1) 用户接口：主要包括 CLI、JDBC/ODBC 客户端和 Web 接口。其中，CLI 为 Shell 命令行；JDBC/ODBC 是 Hive 的 Java 接口实现，与传统数据库 JDBC 类似；Web 接口通过浏览器访问 Hive。

(2) 元数据库：Hive 将元数据存储在数据库中(MySQL 或者 Derby)。Hive 中的元数据包括表的名字、表的列和分区及其属性、表的属性(是否为外部表等)、表的数据所在目录等。

(3) Thrift 服务器：允许客户端使用包括 Java 或其他很多种语言，通过编程的方式远程访问 Hive。

(4) 解释器、编译器、优化器、执行器：完成 HQL 查询语句从词法分析、语法分析、编译、优化以及查询计划的生成。生成的查询计划存储在 HDFS 中，并在随后调用执行 MapReduce。

5.1.3 Hive 的优缺点

1. Hive 的优点

（1）适合大数据的批量处理，解决了传统关系数据库在大数据处理上的瓶颈。

（2）Hive 构建在 Hadoop 之上，充分利用了集群的存储资源、计算资源，最终实现并行计算。

（3）Hive 学习使用成本低。Hive 支持标准的 SQL 语法，免去了编写 MapReduce 程序的过程，减少了开发成本。

（4）具有良好的扩展性，且能够实现和其他组件的结合使用。

2. Hive 的缺点

（1）HQL 的表达能力依然有限。由于本身 SQL 的不足，不支持迭代计算，有些复杂的运算用 HQL 不易表达，还需要单独编写 MapReduce 来实现。

（2）Hive 的运行效率低、延迟高。Hive 是转换成 MapReduce 任务来进行数据分析，MapReduce 是离线计算，所以 Hive 的运行效率也很低，而且是高延迟。

（3）Hive 调优比较困难。由于 Hive 是构建在 Hadoop 之上的，Hive 的调优还要考虑 MapReduce 层面，因此 Hive 的整体调优比较困难。

5.2 Hive 的安装

5.2.1 安装 MySQL

（1）下载 MySQL 的 yum 源，在 CentOS 命令行中输入如下命令：

```
wget http://dev.mysql.com/get/mysql57-community-release-el7-7.noarch.rpm
```

（2）查看下载源中包含的 rpm 包，在 CentOS 命令行中输入如下命令：

```
rpm -qpl mysql57-community-release-el7-7.noarch.rpm
```

（3）安装 MySQL，在 CentOS 命令行中输入如下命令：

```
yum install -y mysql-community-server
```

（4）启动 mysqld 服务，在 CentOS 命令行中输入如下命令：

```
systemctl start mysqld.service
```

（5）查找 MySQL 初始密码，在 CentOS 命令行中输入如下命令：

```
grep "password"  /var/log/mysqld.log
```

运行结果如下所示：

```
[root@bigdata02 bin]# grep "password" /var/log/mysqld.log
2020-06-06T01:40:43.399513Z 1 [Note] A temporary password is generated for root@localhost: Kel0d6tmrT/b
```

(6) 登录 MySQL,在 CentOS 命令行中输入 mysql -uroot -p,输入第(5)步查找的初始密码。

```
[root@bigdata02 bin]# mysql -uroot -p
Enter password:
```

(7) 设置安全级别,MySQL 命令如下：

set global validate_password_policy = 0;

(8) 设置密码长度,MySQL 命令如下：

set global validate_password_length = 4;

(9) 设置密码,MySQL 命令如下：

set password = password('1234');

(10) 创建数据库,MySQL 命令如下：

create database hive;

运行结果如下所示：

```
mysql> set global validate_password_policy=0;
Query OK, 0 rows affected (0.01 sec)

mysql> set global validate_password_length=4;
Query OK, 0 rows affected (0.00 sec)

mysql> set password=password('1234');
Query OK, 0 rows affected, 1 warning (0.00 sec)

mysql> show databases;
+--------------------+
| Database           |
+--------------------+
| information_schema |
| hive               |
| metastore          |
| mysql              |
| performance_schema |
| sys                |
+--------------------+
6 rows in set (0.00 sec)
```

(11) 创建 MySQL 用户,名称为 hadoop,设置用户密码,MySQL 命令如下：

grant all on *.* to hadoop@'%' identified by "hadoop";
grant all on *.* to hadoop@'localhost' identified by 'hadoop';
grant all on *.* to hadoop@'bigdata02' identified by 'hadoop';
flush privileges;

运行结果如下所示：

```
mysql> grant all on *.* to hadoop@'%' identified by "hadoop";
Query OK, 0 rows affected, 1 warning (0.00 sec)

mysql> grant all on *.* to hadoop@'localhost' identified by 'hadoop';
Query OK, 0 rows affected, 1 warning (0.00 sec)

mysql> grant all on *.* to hadoop@'bigdata02' identified by 'hadoop';
Query OK, 0 rows affected, 1 warning (0.00 sec)

mysql> flush privileges;
Query OK, 0 rows affected (0.00 sec)
```

5.2.2 安装 Hive

(1) 下载安装包,下载镜像为 http://mirror.bit.edu.cn/apache/hive/hive-2.3.7/apache-hive-2.3.7-bin.tar.gz。

将安装包上传到服务器,并进行解压,其中 hadoop 是 Hive 的安装目录,命令如下:

```
tar -zxvf apache-hive-2.3.7-bin.tar.gz -C /hadoop/
```

(2) 配置 Hive 环境变量,在 etc/profile 文件中加入环境变量,内容如下:

```
export HIVE_HOME=/hadoop/apache-hive-2.3.7-bin/
export PATH=$HIVE_HOME/bin:$PATH
```

配置完成并保存后,刷新配置文件,执行命令 source /etc/profile。

(3) 在 Hadoop 下创建 hive 文件夹,在 CentOS 命令行上执行以下 Hadoop 命令:

```
hadoop fs -mkdir -p /hive/tmp
hadoop fs -mkdir -p /hive/warehouse
hadoop fs -chmod g+w /hive/tmp
hadoop fs -chmod g+w /hive/warehouse
hadoop fs -chmod -R 777 /user/hive/tmp
```

(4) 修改 Hive 配置文件,在 CentOS 命令行上执行以下命令:

```
cd /hadoop/apache-hive-2.3.7-bin/etc
cp hive-env.sh.template hive-env.sh
cp hive-default.xml.template hive-site.xml
cp hive-log4j2.properties.template hive-log4j2.properties
cp hive-exec-log4j2.properties.template hive-exec-log4j2.properties
```

在 hive-env.sh 文件中加入 JAVA_HOME、HADOOP_HOME 配置,内容如下:

```
export JAVA_HOME=/usr/java/jdk1.8.0_161
export HADOOP_HOME=/hadoop/hadoop-2.7.7
```

在 hive-site.xml 文件中加入以下配置,注意 XML 文件中的 ConnectionUserName 和 ConnectionPassword 配置的是 MySQL 的用户名和密码。

```
<?xml version="1.0" encoding="UTF-8" standalone="no"?>
<configuration>
    <property>
        <name>hive.exec.scratchdir</name>
        <value>/hive/tmp</value>
    </property>
    <property>
        <name>hive.metastore.warehouse.dir</name>
        <value>/hive/warehouse</value>
    </property>
    <property>
        <name>hive.querylog.location</name>
```

```xml
            <value>hive/log</value>
        </property>
    <!-- 配置MySQL数据库连接信息 -->
        <property>
            <name>javax.jdo.option.ConnectionURL</name>
            <value>jdbc:mysql://localhost:3306/metastore?createDatabaseIfNotExist=true&amp;characterEncoding=UTF-8&amp;useSSL=false</value>
        </property>
        <property>
            <name>javax.jdo.option.ConnectionDriverName</name>
            <value>com.mysql.jdbc.Driver</value>
        </property>
        <property>
            <name>javax.jdo.option.ConnectionUserName</name>
            <value>hadoop</value>
        </property>
        <property>
            <name>javax.jdo.option.ConnectionPassword</name>
            <value>hadoop</value>
        </property>
</configuration>
```

（5）初始化 Hive，在 CentOS 命令行上执行以下命令：

```
./schematool -dbType mysql -initSchema hive hive
```

（6）下载 mysql-connector-java-5.1.46.jar，并复制到 Hive 安装目录的 lib 目录下。

（7）启动 Hive，在 CentOS 命令行上执行命令 hive，运行结果如下所示。

```
./hive
SLF4J: Class path contains multiple SLF4J bindings.
SLF4J: Found binding in [jar:file:/hadoop/apache-hive-2.3.7-bin/lib/log4j-slf4j-impl-2.6.2.jar!/org/slf4j/impl/StaticLoggerBinder.class]
SLF4J: Found binding in [jar:file:/hadoop/hadoop-2.7.7/share/hadoop/common/lib/slf4j-log4j12-1.7.10.jar!/org/slf4j/impl/StaticLoggerBinder.class]
SLF4J: See http://www.slf4j.org/codes.html#multiple_bindings for an explanation.
SLF4J: Actual binding is of type [org.apache.logging.slf4j.Log4jLoggerFactory]

Logging initialized using configuration in file:/hadoop/apache-hive-2.3.7-bin/conf/hive-log4j2.properties Async: true
Hive-on-MR is deprecated in Hive 2 and may not be available in the future versions. Consider using a different execution engine (i.e. spark, tez) or using Hive 1.X releases.
hive>
```

5.3 Hive 数据库相关操作

Hive 是基于 Hadoop 构建的一套数据仓库分析系统，它提供了丰富的 SQL 查询方式来分析存储在 Hadoop 分布式文件系统中的数据，可以将结构化的数据文件映射为一张数据库表，并提供完整的 SQL 查询功能，可以将 SQL 语句转换为 MapReduce 任务进行运行，通过自己的 SQL 去查询分析需要的内容。这套 SQL 简称 HQL。HQL 可以使不熟悉 MapReduce 的用户很方便地利用 SQL 查询、汇总和分析数据。而 MapReduce 开发人员可以将自己编写的 Mapper 和 Reducer 作为插件来支持 Hive 做更复杂的数据分析。Hive 中的数据库和常见的关系数据库中的数据库的作用几乎是一样的，在生产环境中,如果表非常多,一般都会用数据库把表组织起来,形成逻辑组,这样可以有效防止大规模集群中表名冲突的问题。Hive 数据库也是用来组织数据表的,它的本质就是数据仓库下的一个目录。

5.3.1 Hive 的数据类型

Hive 的内置数据类型可以分为两大类,分别是基础数据类型和复杂数据类型,Hive 的基础数据类型如表 5-2 所示,Hive 的复杂数据类型如表 5-3 所示。

表 5-2 Hive 的基础数据类型

数 据 类 型	描　　述
tinyint	1 字节的有符号整数
smallint	2 字节有符号整数
int	4 字节有符号整数
bigint	8 字节有符号整数
float	4 字节单精度浮点数
double	8 字节双精度浮点数
double precision	double 的别名,从 Hive 2.2.0 开始提供
decimal	任意精度的带符号小数
numeric	同样是 decimal,从 Hive 3.0 开始
timestamp	精度到纳秒的时间戳
date	以年/月/日形式描述的日期
interval	表示时间间隔
string	字符串
varchar	同 string,字符串长度不固定
char	固定长度的字符串
boolean	布尔型

表 5-3 Hive 的复杂数据类型

数 据 类 型	描　　述
array	一组有序字段,字段类型必须相同
map	一组无序键值对。键的类型必须是原子类型,值可以是任意类型,同一个映射的键的类型必须相同,值的类型也必须相同
struct	一组命名的字段,字段的类型可以不同

5.3.2 Hive 基础 SQL 语法

1. DDL 操作

DDL(Data Definition Language)是数据定义语言,与关系数据库操作相似。
1) 创建数据库
例如,创建一个名为 bigdata 的数据库,DDL 语句为"create database bigdata;",运行

结果如下：

```
hive> create database bigdata;
OK
Time taken: 0.122 seconds
```

显示数据库命令，DDL 语句为"show database;"，运行结果如下：

```
hive> show databases;
OK
default
hhao
Time taken: 3.343 seconds, Fetched: 2 row(s)
```

使用数据库，例如使用一个名为 bigdata 的数据库，DDL 语句为"use bigdata;"，运行结果如下：

```
hive> use bigdata;
OK
Time taken: 0.013 seconds
```

2) 创建表

Hive 在创建表时默认创建内部表，将数据移动到数据仓库指向的路径，而创建外部表（需要加关键字 external），仅记录数据所在的路径，不对数据的位置做任何改变；Hive 删除表时，内部表的元数据和数据会被一起删除，而外部表只删除元数据，不删除数据。创建表的语法如下：

    create table ♯创建一个指定名字的表.如果相同名字的表已经存在,则抛出异常
                ♯用户可以用 if not exist 选项来忽略这个异常

- external：关键字，可以让用户创建一个外部表。
- comment：可以为表与字段增加描述。
- partitioned by：分区。
- clustered by：数据汇总。
- sorted by：按某列排序。
- buckets：分桶，设置词句分桶才有效，如"set hive.enforce.bucketing=true;"。
- row format delimited：按分割格式读取文件。
  [fields terminated by char]：每个列字段通过什么分割。
  [collection items terminated by char]：每个键值之间分割符。
  [map keys terminated by char]：每个键值对分割符。
  [lines terminated by char]：每行之间通过什么分割。
- stored as：存储为不同文件格式。
  [textfile]：文本文件。
  [sequencefile]：序列化文件。
  [rcfile]：面向列的数据格式文件。
  [inputformat input_format_classname outputformat output_format_classname]：自定义的输入输出流。
- location：在建表的同时指定一个指向实际数据的路径。

(1) 创建内部表。例如,创建一个名为 student 的内部表,DDL 语句为:

```
create table if not exists student(
id int,
name string,
birthday timestamp)
row format delimited
fields terminated by '\t'
lines terminated by '\n'
stored as textfile
location '/hive/warehouse/ student';
```

运行结果如下:

```
hive> create table if not exists student(
    > id int,
    > name string,
    > birthday timestamp)
    > row format delimited
    > fields terminated by '\t'
    > lines terminated by '\n'
    > stored as textfile
    > location '/hive/warehouse/ student';
OK
Time taken: 0.267 seconds
```

查看当前数据库中表的命令为 show tables,运行结果如下:

```
hive> show tables;
OK
student
Time taken: 0.023 seconds, Fetched: 1 row(s)
```

查看数据库中的表,使用命令 show tables in bigdata。

查看数据表的结构信息使用命令 desc,比如查看 student 表结构信息的命令为 desc student,运行结果如下:

```
hive> desc student;
OK
id                      int
name                    string
birthday                timestamp
Time taken: 0.051 seconds, Fetched: 3 row(s)
```

通过复制另一张表的表结构来创建表。要注意使用这种复制方式创建的表只复制表结构,不复制表中的数据,比如参照 bigdata 数据库中的 student 表的结构创建表名为 student_copy 的数据表,DDL 语句为"create table if not exists student_copy like student;",运行结果如下:

```
hive> create table if not exists student_copy like student;
OK
Time taken: 0.315 seconds
```

(2) 创建外部表,例如创建一个名为 student_external 的外部表,DDL 语句为:

```
create external table if not exists student_external(
id int,
name string,
birthday timestamp)
row format delimited
fields terminated by '\t'
lines terminated by '\n'
```

```
stored as textfile
location '/hive/warehouse/external';
```

运行结果如下：

```
hive> create external table if not exists student_external(
    > id int ,
    > name string,
    > birthday timestamp)
    > row format delimited
    > fields terminated by '\t'
    > lines terminated by '\n'
    > stored as textfile
    > location '/hive/warehouse/external';
OK
Time taken: 0.066 seconds
```

(3) 创建分区表。例如创建一个名为 teacher_partition 的分区表，并设置成动态建立分区，DDL 语句为：

```
create   table teacher_partition
 (id string,
  name string)
  partitioned by (country string ,state string);
set hive.exec.dynamic.partition = true;                    #开启动态分区,默认是 false
     set hive.exec.dynamic.partition.mode = nonstrict;     #开启允许所有分区都是动态的,否
                                                           #则必须要有静态分区才能使用
set hive.exec.max.dynamic.partitions.pernode = 1000;       #动态分区最大数量
```

运行结果如下：

```
hive> create   table teacher_partition
    > (id string,
    > name string)
    > partitioned by (country string ,state string);
OK
Time taken: 0.056 seconds
hive> set hive.exec.dynamic.partition=true;
hive> set hive.exec.dynamic.partition.mode=nonstrict;
hive> set hive.exec.max.dynamic.partitions.pernode=1000;
hive>
```

(4) 创建桶表。在 Hive 分区表中，分区中的数据量过于庞大时，建议使用桶表。桶表是对某一列数据进行 Hash 取值以将数据打散，然后放到不同文件中存储。在分桶时，对指定字段的值进行 Hash 运算得到 Hash 值，并使用 Hash 值除以桶的个数再取余运算得到的值进行分桶，保证每个桶中有数据但每个桶中的数据不一定相等。做 Hash 运算时，Hash() 函数的选择取决于分桶字段的数据类型。桶表按某一列的值将记录进行分桶存放，即分文件存放，即将大表分解成一系列小表，涉及 Join 操作时，可以在桶与桶间关联即可，这样便减小了 Join 的数据量，提高了执行效率。

创建一个名为 teacher_bucket 的桶表，按照 id 分为 4 个桶，DDL 语句为：

```
create table teacher_bucket(
 id string,
 name string,
 country string,
 state string)
 clustered by(id) into 4 buckets;
```

运行结果如下：

```
hive> use bigdata;
OK
Time taken: 3.3 seconds
hive> create table teacher_bucket(
    > id string,
    > name string,
    > country string,
    > state string)
    > clustered by(id) into 4 buckets;
OK
Time taken: 0.488 seconds
hive> show tables;
OK
sp_trans
student
student_external
teacher
teacher_bucket
teacher_partition
Time taken: 0.08 seconds, Fetched: 6 row(s)
```

3) 修改表

在 Hive 中,使用 alter table 子句来修改表的属性,实际上是修改表的元数据,而不会修改表中实际的数据。

(1) 对表重命名。例如,通过 rename to 命令将表 student_copy 表重命名为 student01,DDL 语句及运行结果如下:

```
hive> alter table student_copy rename to student01;
OK
Time taken: 0.081 seconds
```

(2) 修改列信息。例如,通过 change column 命令对表 student01 的 id 字段重命名,将 id 字段重命名为 uid,DDL 语句及运行结果如下:

```
hive> alter table student01
    > change column id uid int;
OK
Time taken: 0.074 seconds
hive> desc student01;
OK
uid                     int
name                    string
birthday                timestamp
Time taken: 0.029 seconds, Fetched: 3 row(s)
```

(3) 修改字段的类型。例如,通过 change column 命令将 uid 的类型由 int 改为 string,DDL 语句及运行结果如下:

```
hive> desc student01;
OK
uid                     int
name                    string
birthday                timestamp
Time taken: 0.029 seconds, Fetched: 3 row(s)
hive> alter table student01
    > change column  uid  uid string;
OK
Time taken: 0.066 seconds
hive> desc student01;
OK
uid                     string
name                    string
birthday                timestamp
Time taken: 0.028 seconds, Fetched: 3 row(s)
```

(4) 增加列。例如,通过 add columns 命令将 student01 表添加 grade 和 class 两个列,DDL 语句为"alter table student01 add columns(grade string,class string);",DDL 语句及运行结果如下:

```
hive> desc student01;
OK
id                      int
name                    string
birthday                timestamp
Time taken: 0.022 seconds, Fetched: 3 row(s)
hive> alter table student01 add columns(grade string,class string);
OK
Time taken: 0.074 seconds
hive> desc student01;
OK
id                      int
name                    string
birthday                timestamp
grade                   string
class                   string
Time taken: 0.022 seconds, Fetched: 5 row(s)
```

(5) 删除或替换列。例如，通过 replace columns 命令删除或替换 student01 表中的 grade 和 class 两个列，DDL 语句及运行结果如下：

```
hive> alter table student01 replace columns(grade int,class string);
OK
Time taken: 0.059 seconds
```

4）删除表。

例如，使用 drop 命令删除 student01 表，DDL 语句及运行结果如下：

```
hive> show tables;
OK
student
student01
Time taken: 0.013 seconds, Fetched: 2 row(s)
hive> drop table student01;
OK
Time taken: 0.113 seconds
hive> show tables;
OK
student
Time taken: 0.011 seconds, Fetched: 1 row(s)
```

2. DML 操作

DML(Data Manipulation Language)即数据操作语言，是用来对 Hive 数据库中的数据进行操作的语言。数据操作主要是如何向表中装载数据和如何将表中的数据导出，主要操作命令有 load、insert 等，insert 语句的语法基本与标准 SQL 相同。

1）装载数据

(1) 从本地路径下一次性装载大量的数据的方式。以上面建立的内部表 teacher 为例，DML 语句为"load data local inpath '/hadoop/data' into table teacher;"。该语句会把本地的/hadoop/data 文件夹下的所有文件都追加到 teacher 表，其实际上就是一个文件的移动。如果加上 local 关键字，Hive 会将本地文件复制一份然后再上传到指定目录，如果不加 local 关键字，Hive 只会将 HDFS 上的数据移动到指定目录。DML 语句及运行结果如下：

```
hive> load data local inpath '/data.txt' into table teacher;
Loading data to table bigdata.teacher
OK
Time taken: 0.3 seconds
hive> select * from teacher;
OK
1       tom     US      CA
2       jack    US      CB
3       mike    CA      BB
4       ariana  CA      BC
Time taken: 0.075 seconds, Fetched: 4 row(s)
```

(2) 覆盖表中已有的记录。需要加上 overwrite 关键字。以 teacher 为例，DML 语句为"load data local inpath '/data.txt' overwrite into table teacher;"。

(3)装载分区表数据。分区表在 DML 命令中需要指定分区,以分区表 teacher_partition 为例,DML 语句为"load data local inpath '/data.txt' overwrite into table teacher_partition partition(country='US',state='CA');",运行结果如下:

```
hive> load data local inpath '/data.txt' overwrite into table teacher_partition partition (country='US',state='CA');
Loading data to table bigdata.teacher_partition partition (country=US, state=CA)
OK
Time taken: 0.778 seconds
```

2)插入数据

(1)通过标准 SQL 插入数据。例如,向 teacher 表插入一条记录,DML 语句为"insert into teacher(id,name,city,state) values(5,'aliy','CA','BD');",运行结果如下:

```
hive> insert into teacher(id,name,city,state) values(5,'aliy','CA','BD');
WARNING: Hive-on-MR is deprecated in Hive 2 and may not be available in the future versions. Consider u
sing a different execution engine (i.e. spark, tez) or using Hive 1.X releases.
Query ID = root_20200614222457_27db22fc-05e7-48d7-be95-4664f696c393
Total jobs = 3
Launching Job 1 out of 3
Number of reduce tasks is set to 0 since there's no reduce operator
Starting Job = job_1591063491031_0025, Tracking URL = http://bigdata01:8088/proxy/application_1591063491031_0025/
Kill Command = /hadoop/hadoop-2.7.7/bin/hadoop job  -kill job_1591063491031_0025
Hadoop job information for Stage-1: number of mappers: 1; number of reducers: 0
2020-06-14 22:25:05,434 Stage-1 map = 0%,  reduce = 0%
2020-06-14 22:25:10,636 Stage-1 map = 100%,  reduce = 0%, Cumulative CPU 1.9 sec
MapReduce Total cumulative CPU time: 1 seconds 900 msec
Ended Job = job_1591063491031_0025
Stage-4 is selected by condition resolver.
Stage-3 is filtered out by condition resolver.
Stage-5 is filtered out by condition resolver.
Moving data to directory hdfs://ns1/hive/warehouse/bigdata.db/teacher/.hive-staging_hive_2020-06-14_22-24-57_330_5921734926122211004-1/-ext-10000
Loading data to table bigdata.teacher
MapReduce Jobs Launched:
Stage-Stage-1: Map: 1   Cumulative CPU: 1.9 sec   HDFS Read: 4265 HDFS Write: 84 SUCCESS
Total MapReduce CPU Time Spent: 1 seconds 900 msec
OK
Time taken: 15.682 seconds
```

(2)通过查询语句向表中插入数据。例如,将 teacher 表的数据插入 teacher01 表中,DML 语句为"insert into teacher01 select * from teacher;",运行结果如下:

```
hive> insert into teacher01 select * from teacher;
WARNING: Hive-on-MR is deprecated in Hive 2 and may not be available in the future versions. Consider u
sing a different execution engine (i.e. spark, tez) or using Hive 1.X releases.
Query ID = root_20200614223104_bbf17a19-d19a-48d0-97df-b2a4b0ab3024
Total jobs = 3
Launching Job 1 out of 3
Number of reduce tasks is set to 0 since there's no reduce operator
Starting Job = job_1591063491031_0027, Tracking URL = http://bigdata01:8088/proxy/application_1591063491031_0027/
Kill Command = /hadoop/hadoop-2.7.7/bin/hadoop job  -kill job_1591063491031_0027
Hadoop job information for Stage-1: number of mappers: 1; number of reducers: 0
2020-06-14 22:31:09,342 Stage-1 map = 0%,  reduce = 0%
2020-06-14 22:31:15,531 Stage-1 map = 100%,  reduce = 0%, Cumulative CPU 1.97 sec
MapReduce Total cumulative CPU time: 1 seconds 970 msec
Ended Job = job_1591063491031_0027
Stage-4 is selected by condition resolver.
Stage-3 is filtered out by condition resolver.
Stage-5 is filtered out by condition resolver.
Moving data to directory hdfs://ns1/hive/warehouse/bigdata.db/teacher01/.hive-staging_hive_2020-06-14_22-31-04_376_5329050658129452336-1/-ext-10000
Loading data to table bigdata.teacher01
MapReduce Jobs Launched:
Stage-Stage-1: Map: 1   Cumulative CPU: 1.97 sec   HDFS Read: 4167 HDFS Write: 140 SUCCESS
Total MapReduce CPU Time Spent: 1 seconds 970 msec
OK
Time taken: 12.455 seconds
```

注意,通过查询语句向表中插入数据时要保证两个表的格式是一致的,这里的一致指的是要保证查询结果的格式和插入表的格式一致。加 overwrite 表示对源表数据进行覆盖。如果是分区表,则必须用 partition 指定分区。

(3)导出数据。使用 insert 子句可以将数据导出到本地(加 local)或 HDFS,例如将 teacher 表中的数据导出到本地的'/hadoop/data'目录下,DML 语句为"insert overwrite local directory '/hadoop/data' select * from teacher;",运行结果如下：

（4）删除、更新数据。Hive 从 0.14 版本开始支持事务和行级更新,但默认是不支持的,需要一些附加配置。要想支持行级 update、delete,需要配置 Hive 支持事务。在此介绍表全删除命令 truncate,DML 语句为"truncate table teacher;",运行结果如下：

3. DQL 操作

DQL(Data Query Language)即数据查询语言,实现数据的简单查询,主要操作命令有 select、where 等。可以在查询时对数据进行排序、分组等操作,如 sort by、order by、group by 等,同时支持嵌套查询,例如执行 DQL 语句对 teacher 表按城市名称进行分组,同时查询出教师在同一城市数量大于 1 的城市。DML 语句为"select count(*) from teacher group by city having count(*)>1;"。

5.4 本章小结

本章主要介绍了 Hive 的架构、安装和基本操作,因为 Hive 的安装需要 MySQL 数据库,因此也相应介绍了 MySQL 的安装。本章的重点是 Hive 的基本操作,主要是 DDL、DML、DQL 基本操作。

第 6 章

HBase分布式数据库

6.1 HBase 概述

6.1.1 HBase 的架构

HBase 是一个分布式的、面向列的开源数据库,该技术来源于 Fay Chang 所撰写的论文《Bigtable:一个结构化数据的分布式存储系统》。就像 Bigtable 利用了谷歌文件系统(File System)所提供的分布式数据存储一样,HBase 在 Hadoop 之上提供了类似于 Bigtable 的能力。HBase 作为 Apache Hadoop 的子项目,不同于一般的关系数据库,它是一个适合于非结构化数据存储的数据库。其实 HBase 是基于列模式的存储结构。HBase 的架构如图 6-1 所示。

1. 客户端

客户端是整个 HBase 系统的入口,可以通过客户端直接操作 HBase。客户端使用 HBase 的 RPC 机制与 HMaster 和 RegionServer 进行通信。对于管理方面的操作,客户端与 HMaster 进行 RPC 通信;对于数据的读写操作,客户端与 HRegionServer 进行 RPC 交互。HBase 有很多个客户端,除了 Java 客户端外,还有 Thrift、Avro、Rest 等客户端模式。

2. Zookeeper

分布式协调服务 Zookeeper 负责管理 HBase 中多个 HMaster 的"选举",保证在任何时候,集群中只有一个 Active HMaster;存储所有 Region 的寻址入口,实时监控 HRegionServer 的上线和下线信息,并实时通知给 HMaster;存储 HBase 的 Schema 和 Table 元数据。

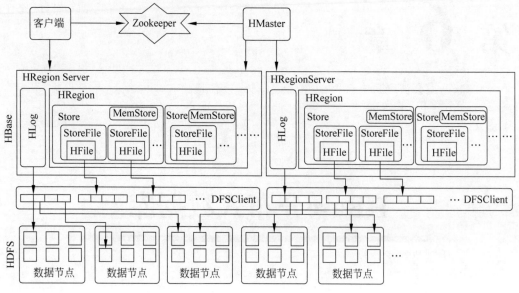

图 6-1　HBase 的架构

3. HMaster

HMaster 是 HBase 的主节点，在 HBase 中可以启动多个 HMaster，通过 Zookeeper 的 Master 选举机制保证总有一个 HMaster 正常运行并提供服务，其他 HMaster 作为备选时刻准备着提供服务。HMaster 主要负责 Table 和 Region 的管理工作。

（1）管理用户对表的增加、删除、修改、查询操作。
（2）管理 HRegionServer 的负载均衡，调整 Region 的分布。
（3）在 Region Split 之后，负责新 Region 的分配。
（4）在 HRegionServer 停机后，负责失效 HRegionServer 的 Region 迁移工作。

4. HRegionServer

HRegionServer 是 HBase 的从节点，主要负责响应用户 I/O 请求，是 HBase 的核心模块。HRegionServer 内部管理了一系列 HRegion 对象，每个 HRegion 对象对应表中的一个 Region。HRegion 由多个 Store 组成，每个 Store 对应表中的一个列族。可以看出，每个列族就是一个集中的存储单元，因此将具备相同 I/O 特性的列放在同一个列族中，能提高读写性能。

5. HRegion

HRegion 即 HBase 表的分片，每个 Region 中保存的是 HBase 表中某段连续的数据。

6. Store

每个 HRegion 包含一或多个 Store。每个 Store 用于管理一个 Region 上的一个列族。

7. MemStore

内存级缓存 MemStore 存储的是用户写入的数据,一旦 MemStore 存储达到阈值,里面存储的数据就会被刷新到新生成的 StoreFile 中(底层是 HFile),该文件是以 HFile 的格式存储到 HDFS 上的。

8. StoreFile

MemStore 中的数据写的文件就是 StoreFile,StoreFile 底层是以 HFile 文件的格式保存在 HDFS 上的。

9. HFile

HBase 中键值对类型的数据均以 HFile 文件格式进行存储。

10. HLog

HLog 为预写日志文件,负责记录 HBase 的修改。当 HBase 读写数据时,数据不是直接写进磁盘,而是会在内存中保留一段时间。

6.1.2 HBase 的特点

(1) 容量巨大。单表可以有百亿行、数百万列。
(2) 无模式。同一个表的不同行可以有截然不同的列。
(3) 面向列。HBase 是面向列的存储和权限控制,并支持列独立索引。
(4) 稀疏性。表可以设计得非常稀疏,值为空的列并不占用存储空间。
(5) 扩展性。HBase 底层文件存储依赖 HDFS,它天生具备可扩展性。
(6) 高可靠性。HBase 提供了预写日志(WAL)和副本(Replication)机制,防止数据丢失。
(7) 高性能。底层的 LSM(Log-Structured Merge Tree)数据结构和 RowKey 有序排列等架构上的独特设计,使得 HBase 具备非常高的写入性能。

6.1.3 HBase 数据存储方式

HBase 表的数据按照行键 RowKey 的字典顺序进行排列,并且切分多个 HRegion 存储,HBase 存储方式示意图如图 6-2 所示。

每个 Region 存储的数据是有限的,如果当 Region 增大到一个阈值(128MB),会被等分切成两个新的 Region,HBase Region 切分示意图如图 6-3 所示。

一个 HRegion Server 上可以存储多个 Region,但是每个 Region 只能被分布到一个 HRegion Server 上。

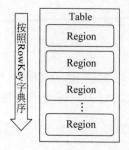

图 6-2　HBase 存储方式示意图

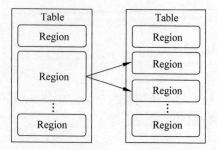

图 6-3　HBase Region 切分示意图

6.1.4　HBase 寻址机制

在介绍 HBase 的寻址之前，先介绍 HBase 的两个特殊的表：-ROOT-表和.META.表。它们是 HBase 的两张内置表，从存储结构和操作方法的角度来说，它们和其他 HBase 的表没有任何区别，可以认为这就是两张普通的表，对普通表的操作对它们都适用。它们与众不同的地方是 HBase 用它们来存储一个重要的系统信息—Region 的分布情况以及每个 Region 的详细信息；并且这两张表的表结构是相同的，但记录的信息不同，-ROOT-表只用来记录.META.表的 Region 信息，而.META.表记录用户表的 Region 信息。

HBase 将-ROOT-表的 RegionServer 的地址存储在 Zookeeper 中，客户端需要先到 Zookeeper 取到-ROOT-表的 RegionServer 的地址，然后根据-ROOT-表的 RegionServer 的地址去访问-ROOT-表。从-ROOT-表中得到.META.表的 RegionServer 地址，访问.META.表，从.META.表中取到用户表的地址信息后去访问用户表。HBase 的寻址流程如图 6-4 所示。

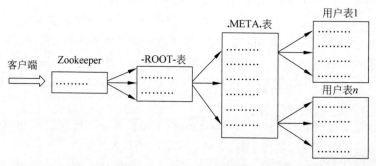

图 6-4　HBase 的寻址流程

假设.META.表被分成了两个 Region，-ROOT-表内容示意图如图 6-5 所示。.META.表内容示意图如图 6-6 所示。

RowKey	Info			Historian
	Regioninfo	Server	Server Startcode	
.META.,Table1,0,12345678,12657843		RS1		
.META.,Table2,30000,12348765,12438675		RS2		

图 6-5　-ROOT-表内容示意图

RowKey	Info			Historian
	Regioninfo	Server	Server Startcode	
Table1,RK0,12345678		RS1		
Table1,RK10000,12345687		RS2		
Table1,RK20000,12346578		RS3		
⋮	⋮	⋮	⋮	⋮
Table2,RK0,12345678		RS1		

图 6-6　META. 表内容示意图

6.2　HBase 的安装

（1）安装 JDK、Hadoop 以及 Zookeeper。本例使用三台虚拟机安装 HBase 集群，虚拟机信息及安装 JDK、Hadoop 和 Zookeeper 的步骤详见第 2 章。

（2）下载 HBase 安装包。这里选择下载的版本是 2.0，下载地址为 http://archive.apache.org/dist/HBase/2.0.0-alpha-2/。

（3）上传并解压 HBase 安装包。将 HBase 安装包上传至 CentOS 操作系统的指定目录并进行解压。

（4）修改配置文件 hbase-env.sh、hbase-site.xml、regionservers，文件路径在 HBase 解压路径的 conf 目录下。

① 配置 hbase-env.sh 文件，在文件中加入以下内容：

```
export JAVA_HOME = 自己的 JAVA_HOME 路径
export HBASE_MANAGES_ZK = false
HBASE_MANAGES_ZK = false   //不使用内部的 Zookeeper,而是使用外部搭建的 Zookeeper 集群
```

② 配置 hbase-site.xml 文件，内容如下：

```
<configuration>
    <!-- HBase 数据目录位置,其中 bigdata01 是 HadoopMaster 机器的机器名 -->
    <property>
        <name>hbase.rootdir</name>
        <value>hdfs://bigdata01:9000/HBase</value>
    </property>
    <!-- 启用分布式集群 -->
    <property>
        <name>hbase.cluster.distributed</name>
        <value>true</value>
    </property>
    <!-- 默认 HMaster HTTP 访问端口 -->
```

```xml
<property>
    <name>hbase.master.info.port</name>
    <value>16010</value>
</property>
<!-- 默认 HRegionServer HTTP 访问端口 -->
<property>
    <name>hbase.regionserver.info.port</name>
    <value>16030</value>
</property>
<!-- 配置独立的 ZK 集群地址 -->
<property>
    <name>hbase.zookeeper.quorum</name>
    <value>bigdata01,bigdata02,bigdata03</value>
</property>
</configuration>
```

③ 配置 regionservers 文件,在文件中加入以下内容:

bigdata01
bigdata02
bigdata03

(5) 配置系统变量,在/etc/profile 文件中加入以下内容:

export HBASE_HOME = HBase 解压路径
export PATH = $HBASE_HOME/bin:$PATH

(6) 系统变量立即生效,在 CentOS 命令行中输入 source /etc/profile 命令。

(7) 复制安装文件至其他两台机器,在 CentOS 命令行中输入如下命令:

scp -r /hadoop/hbase-2.0.0-alpha2/ root@bigdata03:/hadoop/
scp -r /hadoophbase-2.0.0-alpha2/ root@bigdata02:/hadoop/

(8) 启动 HBase,在 bigdata01 机器中启动,在 CentOS 命令行中输入 start-hbase.sh(必须先启动 Hadoop 与 Zookeeper,否则启动失败)。

(9) 查看是否启动成功,在 CentOS 命令行中输入 jps 命令,运行结果如下:

```
[root@bigdata01 bin]# jps
32240 DFSZKFailoverController
11782 HMaster
12134 Jps
31687 NameNode
32023 JournalNode
32503 NodeManager
31304 QuorumPeerMain
32376 ResourceManager
11931 HRegionServer
31805 DataNode
```

(10) 在浏览器上输入 IP:端口号,例如 http://172.16.106.69:16010/,HBase Web 管理界面如图 6-7 所示。

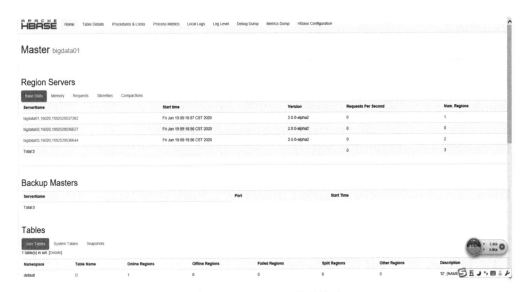

图 6-7　HBase Web 管理界面

6.3　HBase 数据模型

HBase 表是一个稀疏、多维度、排序的映射表，表的索引是行键、列族、列限定符和时间戳。表在水平方向由一个或者多个列族组成，一个列族中可以包含任意多个列，列族支持动态扩展，可以很轻松地添加一个列族或列，无须预先定义列的数量以及数据类型，所有列均以字符串形式存储。用户需要自行进行数据类型转换。HBase 中执行更新操作时，并不会删除数据旧的版本，而是生成一个新的版本，旧的版本仍然保留（这是和 HDFS 只允许追加不允许修改的特性相关）。

（1）表。HBase 采用表来组织数据，表由行和列组成，列划分为若干个列族。

（2）行键。每个 HBase 表都由若干行组成，每个行由行键（Row Key）来标识。

（3）列族。一个 HBase 表被分组成许多列族（Column Family）的集合，它是基本的访问控制单元。

（4）列限定符。列族中的数据通过列限定符（或列）来定位。

（5）单元格。在 HBase 表中，通过行键、列族和列限定符确定一个单元格，单元格中存储的数据没有数据类型，总被视为字节数组。

（6）时间戳。每个单元格都保存着同一份数据的多个版本，这些版本采用时间戳进行索引。

HBase 表数据模型示意图如图 6-8 所示。

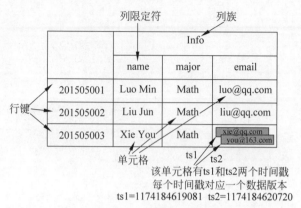

图 6-8　HBase 表数据模型示意图

6.4　HBase 的 Shell 操作

HBase 的 Shell 提供大量操作 HBase 的命令，通过 Shell 命令很方便地操作 HBase 数据库，例如创建、删除及修改表，向表中添加数据，列出表中的相关信息等操作。当使用 Shell 命令行操作 HBase 时，需要进入 HBase Shell 交互界面，执行 ./hbase Shell 命令进入 HBase Shell 交互界面。运行结果如下：

```
[root@bigdata01 bin]# ./hbase shell
SLF4J: Class path contains multiple SLF4J bindings.
SLF4J: Found binding in [jar:file:/hadoop/hbase-2.0.0-alpha2/lib/slf4j-log4j12-1.7.10.jar!/org/s
pl/StaticLoggerBinder.class]
SLF4J: Found binding in [jar:file:/hadoop/hadoop-2.7.7/share/hadoop/common/lib/slf4j-log4j12-1.7
!/org/slf4j/impl/StaticLoggerBinder.class]
SLF4J: See http://www.slf4j.org/codes.html#multiple_bindings for an explanation.
SLF4J: Actual binding is of type [org.slf4j.impl.Log4jLoggerFactory]
HBase Shell
Use "help" to get list of supported commands.
Use "exit" to quit this interactive shell.
Version 2.0.0-alpha2, r7149f999786b6fd5a3fc1f7aec1214afb738925e, Wed Aug 16 10:30:21 PDT 2017
Took 0.0030 seconds
hbase(main):001:0>
```

HBase Shell 交互界面中，可通过一系列 Shell 命令操作 HBase，HBase 常见的 Shell 命令如表 6-1 所示。

表 6-1　HBase 常见的 Shell 命令

命令名称	相关说明
create	创建表
put	插入或更新数据
scan	扫描表并返回表的所有数据
describe	查看表的结构
get	获取指定行或列的数据
count	统计表中数据的行数
delete	删除指定行或者列的数据
deleteall	删除整个行或列的数据
truncate	删除整个表中的数据，但是结构还在
drop	删除整个表，数据和结构都删除

下面举例说明 HBase 常见的 Shell 命令。

(1) 显示 HBase 中的表。HBase 命令及运行结果如下：

```
hbase(main):001:0> list
TABLE
t2
1 row(s)
Took 0.3780 seconds
```

(2) 创建表 user，包含 info、data 两个列族。HBase 命令及运行结果如下：

```
hbase(main):002:0> create 'user', 'info', 'data'
Created table user
Took 1.3458 seconds
```

(3) 向 user 表中插入数据。HBase 命令及运行结果如下：

```
hbase(main):012:0> put 'user', 'rk0001', 'info:name', 'zhangsan'
Took 0.0088 seconds
hbase(main):013:0> put 'user', 'rk0001', 'info:gender', 'female'
Took 0.0049 seconds
hbase(main):014:0> put 'user', 'rk0001', 'info:age', 20
Took 0.0102 seconds
hbase(main):015:0> put 'user', 'rk0001', 'data:pic', 'picture'
Took 0.0055 seconds
```

(4) 获取 user 表中 RowKey 为 rk0001 的所有信息。HBase 命令及运行结果如下：

```
hbase(main):014:0> get 'user', 'rk0001'
COLUMN                CELL
 data:pic             timestamp=1592534367612, value=picture
 info:age             timestamp=1592534147984, value=20
 info:gender          timestamp=1592534113227, value=female
 info:name            timestamp=1592534084941, value=zhangsan
1 row(s)
```

(5) 获取 user 表中 RowKey 为 rk0001 和 info 列族的所有信息。HBase 命令及运行结果如下：

```
hbase(main):015:0> get 'user', 'rk0001', 'info'
COLUMN                CELL
 info:age             timestamp=1592534147984, value=20
 info:gender          timestamp=1592534113227, value=female
 info:name            timestamp=1592534084941, value=zhangsan
1 row(s)
```

(6) 获取 user 表中 RowKey 为 rk0001，info 列族的 name、age 列标识符的信息。HBase 命令及运行结果如下：

```
hbase(main):016:0> get 'user', 'rk0001', 'info:name', 'info:age'
COLUMN                CELL
 info:age             timestamp=1592534147984, value=20
 info:name            timestamp=1592534084941, value=zhangsan
1 row(s)
```

(7) 获取 user 表中 RowKey 为 rk0001，info、data 列族的信息。HBase 命令及运行结果如下：

```
hbase(main):017:0> get 'user', 'rk0001', 'info', 'data'
COLUMN                CELL
 data:pic             timestamp=1592534367612, value=picture
 info:age             timestamp=1592534147984, value=20
 info:gender          timestamp=1592534113227, value=female
 info:name            timestamp=1592534084941, value=zhangsan
1 row(s)
Took 0.0164 seconds
hbase(main):018:0> get 'user', 'rk0001', {COLUMN => ['info', 'data']}
COLUMN                CELL
 data:pic             timestamp=1592534367612, value=picture
 info:age             timestamp=1592534147984, value=20
 info:gender          timestamp=1592534113227, value=female
 info:name            timestamp=1592534084941, value=zhangsan
1 row(s)
Took 0.0157 seconds
hbase(main):019:0> get 'user', 'rk0001', {COLUMN => ['info:name', 'data:pic']}
COLUMN                CELL
 data:pic             timestamp=1592534367612, value=picture
 info:name            timestamp=1592534084941, value=zhangsan
1 row(s)
Took 0.0145 seconds
```

（8）获取 user 表中 RowKey 为 rk0001，列族为 info，版本号最新 3 个的信息。HBase 命令及运行结果如下：

```
hbase(main):011:0> alter 'user', {NAME => 'info', VERSIONS => 3}
Updating all regions with the new schema...
1/1 regions updated.
Done.
Took 2.2442 seconds
```

HBase 2.0 默认 VERSIONS 为 1，也就是说，默认情况只会存取一个版本的列数据。需要使用 alter 命令修改表的版本号。

（9）向 user 表 info:name 列插入数据。HBase 命令及运行结果如下：

```
hbase(main):016:0> put 'user', 'rk0001', 'info:name', 'zhangsan1'
Took 0.0038 seconds
```

（10）查看 user 表 info:name 列历史版本。HBase 命令及运行结果如下：

```
hbase(main):017:0> get 'user', 'rk0001', {COLUMN => 'info:name', VERSIONS => 3}
COLUMN                CELL
 info:name            timestamp=1592536721505, value=zhangsan1
 info:name            timestamp=1592536680393, value=zhangsan
1 row(s)
```

（11）获取 user 表中 RowKey 为 rk0001、列标识符中含有 a 的信息。HBase 命令及运行结果如下：

```
hbase(main):018:0> get 'user', 'rk0001', {FILTER => "(QualifierFilter(=,'substring:a'))"}
COLUMN                CELL
 info:age             timestamp=1592536699592, value=20
 info:name            timestamp=1592536721505, value=zhangsan1
1 row(s)
Took 0.0425 seconds
```

（12）利用 scan 命令查询 user 表中的所有信息。

① 查询 user 表中的所有信息。HBase 命令及运行结果如下：

```
hbase(main):019:0> scan 'user'
ROW                   COLUMN+CELL
 rk0001               column=data:pic, timestamp=1592536711273, value=picture
 rk0001               column=info:age, timestamp=1592536699592, value=20
 rk0001               column=info:gender, timestamp=1592536689035, value=female
 rk0001               column=info:name, timestamp=1592536721505, value=zhangsan1
1 row(s)
Took 0.0400 seconds
```

② 查询 user 表中列族为 info、列标识符为 name 的信息。HBase 命令及运行结果如下：

```
hbase(main):001:0> scan 'user', {COLUMNS => 'info:name'}
ROW                   COLUMN+CELL
 rk0001               column=info:name, timestamp=1592536721505, value=zhangsan1
1 row(s)
Took 0.4020 seconds
```

③ 查询 user 表中列族为 info、RowKey 范围为（rk0001，rk0003）的数据。HBase 命令及运行结果如下：

```
hbase(main):001:0> scan 'user', {COLUMNS => 'info', STARTROW => 'rk0001', ENDROW => 'rk0003'}
ROW                   COLUMN+CELL
 rk0001               column=info:age, timestamp=1592536699592, value=20
 rk0001               column=info:gender, timestamp=1592536689035, value=female
 rk0001               column=info:name, timestamp=1592536721505, value=zhangsan1
1 row(s)
Took 0.4530 seconds
```

④ 查询 user 表中 RowKey 以 rk 字符开头的数据。HBase 命令及运行结果如下：

```
hbase(main):002:0> scan 'user',{FILTER=>"PrefixFilter('rk')"}
ROW                   COLUMN+CELL
 rk0001               column=data:pic, timestamp=1592536711273, value=picture
 rk0001               column=info:age, timestamp=1592536699592, value=20
 rk0001               column=info:gender, timestamp=1592536689035, value=female
 rk0001               column=info:name, timestamp=1592536721505, value=zhangsan1
1 row(s)
```

⑤ 查询 user 表中指定时间范围的数据。HBase 命令及运行结果如下：

```
hbase(main):002:0> scan 'user', {TIMERANGE => [1592536711273, 1592536721505]}
ROW                    COLUMN+CELL
 rk0001                column=data:pic, timestamp=1592536711273, value=picture
1 row(s)
Took 0.0318 seconds
```

（13）删除数据。

① 删除 user 表中 Row Key 为 rk0001、列标识符为 info：name 的数据。HBase 命令及运行结果如下：

```
hbase(main):003:0> delete 'user', 'rk0001', 'info:name'
Took 0.0717 seconds
hbase(main):004:0> scan 'user'
ROW                    COLUMN+CELL
 rk0001                column=data:pic, timestamp=1592536711273, value=picture
 rk0001                column=info:age, timestamp=1592536699592, value=20
 rk0001                column=info:gender, timestamp=1592536689035, value=female
1 row(s)
Took 0.0111 seconds
```

② 清空 user 表数据。HBase 命令及运行结果如下：

```
hbase(main):005:0> truncate 'user'
Truncating 'user' table (it may take a while):
Disabling table...
Truncating table...
Took 1.6295 seconds
hbase(main):006:0> scan 'user'
ROW                    COLUMN+CELL
0 row(s)
Took 0.1996 seconds
```

（14）修改表结构。

① 添加 user 表 f1 列族。HBase 命令及运行结果如下：

```
hbase(main):007:0> alter 'user', NAME => 'f1'
Updating all regions with the new schema...
1/1 regions updated.
Done.
Took 3.3347 seconds
```

② 删除 user 表 f1 列族。HBase 命令及运行结果如下：

```
hbase(main):009:0> alter 'user', 'delete' => 'f1'
Updating all regions with the new schema...
1/1 regions updated.
Done.
Took 2.2443 seconds
```

③ 删除 user 表（先停用表）。HBase 命令及运行结果如下：

```
hbase(main):011:0> disable 'user'
Took 0.4296 seconds
hbase(main):012:0> drop 'user'
Took 0.2410 seconds
hbase(main):013:0>
```

6.5 HBase 常用的 Java API 及示例程序

6.5.1 HBase 常用的 Java API

HBase 是使用 Java 语言开发的，它对外提供了 Java API 的接口。HBase 常用的 Java API 如表 6-2 所示。

表 6-2　HBase 常用的 Java API

类或接口名称	相关说明
Admin	一个类,用于建立客户端和 HBase 数据库的连接
HBaseConfiguration	一个类,用于将 HBase 配置添加到配置文件中
HTableDescriptor	一个接口,用于描述表的信息
HColumnDescriptor	一个类,用于描述列族的信息
Table	一个接口,用于实现 HBase 表的通信
Put	一个类,用于插入数据操作
Get	一个类,用于查询单条记录
Delete	一个类,用于删除数据
Scan	一个类,用于查询所有记录
Result	一个类,用于查询返回的单条记录结果

6.5.2　程序示例

该示例主要功能是利用 HBase 常用的 Java API 进行表的创建、插入、删除等操作,主要步骤如下:

(1) 新建工程。使用 IntelliJ IDEA 新建一个 MAVEN 工程。

(2) 导入依赖包,pom.xml 文件配置内容如下:

```xml
<!-- 单元测试依赖 -->
<dependencies>
    <dependency>
        <groupId>junit</groupId>
        <artifactId>junit</artifactId>
        <version>4.12</version>
    </dependency>
    <!-- HBase 客户端依赖 -->
    <dependency>
        <groupId>org.apache.hbase</groupId>
        <artifactId>hbase-client</artifactId>
        <version>2.0.0</version>
    </dependency>
    <!-- HBase 核心依赖 -->
    <dependency>
        <groupId>org.apache.hbase</groupId>
        <artifactId>hbase-common</artifactId>
        <version>2.0.0</version>
    </dependency>
</dependencies>
```

(3) 新建测试类 HbaseApiDemo,代码及说明如下:

```java
package com.hadoop.hbase;
import org.apache.hadoop.conf.Configuration;
import org.apache.hadoop.hbase.*;
```

```java
import org.apache.hadoop.hbase.client.*;
import org.apache.hadoop.hbase.util.Bytes;
import org.junit.Before;
import org.junit.Test;
import java.io.IOException;
import java.util.ArrayList;
import java.util.Iterator;
import java.util.List;
public class HbaseApiDemo{
    //初始化 Configuration 对象
    private Configuration conf = null;
    //初始化连接
    private Connection conn = null;
    @Before
    public void init() throws Exception{
        //获取 Configuration 对象
        conf = HBaseConfiguration.create();
        //设置 Zookeeper 集群地址

        conf.set("hbase.zookeeper.quorum","bigdata01:2181,bigdata02:2181,bigdata03:2181");
        //获取连接
        conn = ConnectionFactory.createConnection(conf);
    }
    //创建表
    @Test
    public void CreateTable() throws Exception{
        try {
            //获取操作对象
            Admin admin = conn.getAdmin();
            //构建一个 t_user 表
            TableDescriptorBuilder t_user =
            TableDescriptorBuilder.newBuilder
            (TableName.valueOf("t_user"));
             //创建列族 info
             ColumnFamilyDescriptor of =
             ColumnFamilyDescriptorBuilder.of("info");
            t_user.setColumnFamily(of);
            //创建列族 data
             ColumnFamilyDescriptor of1 =
             ColumnFamilyDescriptorBuilder.of("data");
            t_user.setColumnFamily(of1);
            //构建
            TableDescriptor build = t_user.build();
            //创建表
            admin.createTable(build);
            //关闭连接
            admin.close();
            conn.close();
        } catch (Exception e) {
```

```java
                    e.printStackTrace();
                }
            
        }
        //插入数据
        @Test
        public void testPut() throws Exception{
            //创建table对象,通过table对象来添加数据
            Table table = conn.getTable(TableName.valueOf("t_user"));
            //创建一个集合,用于存放put对象
            ArrayList<Put> puts = new ArrayList<Put>();
            //构建put对象(kv形式),并指定其行键
            Put put01 = new Put(Bytes.toBytes("rk002"));
            put01.addColumn(Bytes.toBytes("info"),
                    Bytes.toBytes("username"),Bytes.toBytes("zhangsan"));
            put01.addColumn(Bytes.toBytes("info"),
                    Bytes.toBytes("password"),Bytes.toBytes("345678"));
            Put put02 = new Put("rk003".getBytes());
            put02.addColumn(Bytes.toBytes("info"),
                    Bytes.toBytes("username"), Bytes.toBytes("lisi"));
            //把所有的put对象添加到一个集合中
            puts.add(put01);
            puts.add(put02);
            //提交所有的插入数据的记录
            table.put(puts);
            //关闭
            table.close();
            conn.close();
        }
        //查询表
        @Test
        public void testGet() throws IOException{
            
            Table table = conn.getTable(TableName.valueOf("t_user"));
            //得到用于扫描region的对象
            Get get = new Get("rk002".getBytes());
            //使用HTable得到resultcanner实现类的对象
            Result result1 = table.get(get);
            List<Cell> cells = result1.listCells();
            for (Cell cell : cells){
                //得到rowkey
                System.out.println("行键:" + Bytes.toString(CellUtil.cloneRow(cell)));
                //得到列族
                System.out.println("列族:" + ytes.toString(CellUtil.cloneFamily(cell)));
                System.out.println("列:" + ytes.toString(CellUtil.cloneQualifier(cell)));
                System.out.println("值:" + Bytes.toString(CellUtil.cloneValue(cell)));
            }
            
        }
        @Test
```

```java
public void testScan() throws Exception{
    //获取table对象
    Table table = conn.getTable(TableName.valueOf("t_user"));
    //获取scan对象
    Scan scan = new Scan();
    //获取查询的数据
    ResultScanner scanner = table.getScanner(scan);
    //获取ResultScanner所有数据,返回迭代器
    Iterator<Result> iter = scanner.iterator();
    //遍历迭代器
    while (iter.hasNext()){
        //获取当前每一行结果数据
        Result result = iter.next();
        //获取当前每一行中所有的cell对象
        List<Cell> cells = result.listCells();
        //迭代所有的cell
        for(Cell c:cells){
            //获取行键
            byte[] rowArray = c.getRowArray();
            //获取列族
            byte[] familyArray = c.getFamilyArray();
            //获取列族下的列名称
            byte[] qualifierArray = c.getQualifierArray();
            //列字段的值
            byte[] valueArray = c.getValueArray();
            //打印rowArray、familyArray、qualifierArray、valueArray
            System.out.println("行键:" + new String(rowArray,c.getRowOffset(),
                    c.getRowLength()));
            System.out.print("列族:" + new String(familyArray,c.getFamilyOffset(),
            c.getFamilyLength()));
            System.out.print(" " + "列:" + new String(qualifierArray,
                    c.getQualifierOffset(),c.getQualifierLength()));
            System.out.println(" " + "值:" + new String(valueArray,
                    c.getValueOffset(), c.getValueLength()));
        }
        System.out.println(" ---------------------- ");
    }
    //关闭
    table.close();
    conn.close();
}
//删除表记录
@Test
public void testDel() throws Exception{
    //获取table对象
    Table table = conn.getTable(TableName.valueOf("t_user"));
    //获取delete对象,需要一个Row Key
```

```java
        Delete delete = new Delete("rk002".getBytes());
        //在 delete 对象中指定要删除的列族-列名称
        delete.addColumn("info".getBytes(),"password".getBytes());
        //执行删除操作
        table.delete(delete);
        //关闭
        table.close();
        conn.close();
    }
    //删除表
    @Test
    public void testDrop() throws Exception{
        //获取一个表的管理器
        Admin admin = conn.getAdmin();
        //删除表时先要确定禁用表
        admin.disableTable(TableName.valueOf("t_user"));
        admin.deleteTable(TableName.valueOf("t_user"));
        //关闭
        admin.close();
        conn.close();
    }
}
```

（4）依次运行创建表、插入数据、查询数据等方法。运行查询数据方法如图 6-9 所示，控制台输出结果如图 6-10 所示。

图 6-9　运行查询方法

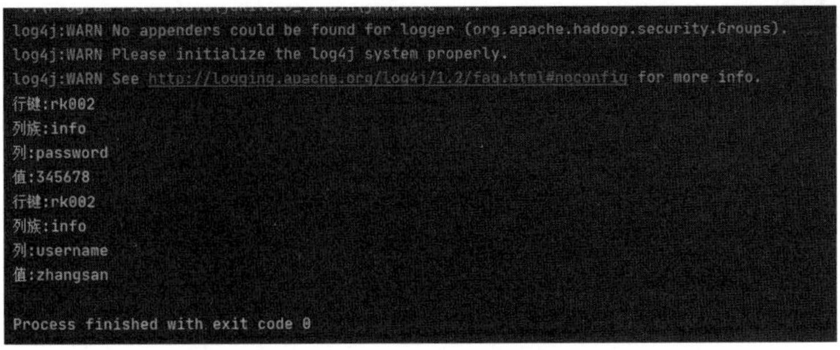

图 6-10 控制台输出结果

6.6 本章小结

本章首先介绍了 HBase 的架构、寻址机制以及安装,然后介绍了 HBase 的 Shell 操作,包括创建表、插入数据、删除等操作。最后介绍了 HBase 常用的 Java API,并且通过程序示例实现了 HBase 的 Java 编程。除了使用 HBase 的 Java API 操作 HBase 数据库外,还可以利用 Phoenix 操作 HBase 数据库,Phoenix 是构建在 HBase 上的一个 SQL 层,是内嵌在 HBase 中的 JDBC 驱动,能够让用户使用标准的 JDBC 来操作 HBase。

第 7 章

Spark 基础

7.1 Spark 概述

Spark 最初由美国加州大学伯克利分校的 AMP(Algorithms, Machines and People) 实验室于 2009 年开发,是基于内存计算的大数据并行计算框架,可用于构建大型的、低延迟的数据分析应用程序。Spark 开始属于研究性项目,其诸多核心理念均源自学术研究论文。2013 年,Spark 加入 Apache 孵化器项目后,开始获得迅猛的发展,如今已成为 Apache 软件基金会最重要的三大分布式计算系统开源项目(即 Hadoop、Spark、Storm) 之一。

Spark 作为大数据计算平台的后起之秀,在 2014 年打破了 Hadoop 保持的基准排序 (Sort Benchmark) 纪录,使用 206 个节点在 23min 内完成了 100TB 数据的排序,而 Hadoop 则是使用 2000 个节点在 72min 内完成同样数据的排序。也就是说,Spark 仅使用了十分之一的计算资源,获得了比 Hadoop 快 2 倍的速度。新纪录的诞生,使得 Spark 获得多方追捧,也表明了 Spark 可以作为一个更加快速、高效的大数据计算平台。

7.1.1 Spark 的主要特点

Spark 的主要特点主要有以下 4 个方面。

(1) 运行速度快。Spark 使用先进的 DAG(Directed Acyclic Graph,有向无环图)执行引擎,以支持循环数据流与内存计算,基于内存的执行速度比 Hadoop MapReduce 快上百倍,基于磁盘的执行速度也比 Hadoop MapReduce 快十倍。

(2) 容易使用。Spark 支持使用 Scala、Java、Python 和 R 语言进行编程,简洁的 API 设计有助于用户轻松构建并行程序,并且可以通过 Spark Shell 进行交互式编程。

（3）通用性。Spark 提供了完整而强大的技术栈，包括 SQL 查询、流式计算、机器学习和图算法组件，这些组件可以无缝整合在同一个应用中，足以应对复杂的计算。

（4）运行模式多样。Spark 可运行于独立的集群模式中，或者运行于 Hadoop 中，也可运行于 Amazon EC2 等云环境中，并且可以访问 HDFS、HBase、Hive 等多种数据源。

Spark 如今已吸引了国内外各大公司的注意，如腾讯、淘宝、百度、亚马逊等公司均不同程度地使用了 Spark 来构建大数据分析应用，并应用到实际的生产环境中。相信在将来，Spark 会在更多的应用场景中发挥重要作用。

Spark 支持使用 Scala、Java、Python 和 R 语言进行编程。由于 Spark 采用 Scala 语言进行开发，因此，建议采用 Scala 语言进行 Spark 应用程序的编写。采用 Scala 语言编写 Spark 应用程序，可以获得最好的性能，和其他语言相比，Scala 主要有以下三个方面的优势。

（1）Java 代码比较烦琐。在大数据应用场景中，不太适合使用 Java，因为完成同样的任务，Scala 只需要一行代码，而 Java 则可能需要 10 行代码。而且，Scala 语言可以支持交互式编程，大大提高了程序开发效率，而 Java 则不支持交互式执行，必须编译以后运行。

（2）Python 语言并发性能不好。在并发性能方面，Scala 要明显优于 Python，而且 Scala 是静态类型，可以在编译阶段就抛出错误，便于开发大型大数据项目。

（3）Scala 兼容 Java。Scala 可以直接使用 Java 中的 Hadoop API 来和 Hadoop 进行交互，但是，Python 与 Hadoop 之间的交互通常都需要第三方库（如 Hadoop）。

7.1.2 Spark 生态系统

Spark 生态系统已经发展成为一个可应用于大规模数据处理的统一分析引擎，它是基于内存计算的大数据并行计算框架，适用于各种各样的分布式平台的系统。在 Spark 生态圈中包含了 Spark SQL、Spark Streaming、MLlib、GraphX 等组件。Spark 生态系统如图 7-1 所示。

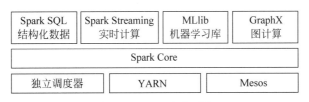

图 7-1 Spark 生态系统

（1）Spark SQL。用来操作结构化数据的核心组件，通过 Spark SQL 可直接查询 Hive、HBase 等多种外部数据源中的数据。Spark SQL 的重要特点是能够统一处理关系表和 RDD。

（2）Spark Streaming。Spark 提供的流式计算框架，支持高吞吐量、可容错处理的实时流式数据处理，其核心原理是将流数据分解成一系列短小的批处理作业。

（3）MLlib。Spark 提供的关于机器学习功能的算法程序库，包括分类、回归、聚类、协同过滤算法等，还提供了模型评估、数据导入等额外的功能。

（4）GraphX。Spark 提供的分布式图处理框架，拥有对图计算和图挖掘算法的 API 接口及丰富的功能和运算符，便于满足分布式图处理的需求，能在海量数据上运行复杂的图算法。

（5）Spark Core。Spark 核心组件，实现了 Spark 的基本功能，包含任务调度、内存管理、错误恢复、与存储系统交互等模块。Spark Core 中还包含对弹性分布式数据集的 API 定义。

（6）独立调度器、YARN、Mesos。集群管理器，负责 Spark 框架高效地在一个到数千个节点之间进行伸缩计算的资源管理。

7.1.3　Spark 相对于 Hadoop MapReduce 的优势

（1）编程方式。Hadoop 的 MapReduce 计算数据时，要转化为 Map 和 Reduce 两个过程，从而难以描述复杂的数据处理过程；而 Spark 的计算模型不局限于 Map 和 Reduce 操作，还提供了多种数据集的操作类型，编程模型比 MapReduce 更加灵活。

（2）数据存储。Hadoop 的 MapReduce 进行计算时，每次产生的中间结果都存储在本地磁盘中；而 Spark 在计算时产生的中间结果存储在内存中。

（3）数据处理。Hadoop 在每次执行数据处理时，都要从磁盘中加载数据，导致磁盘 I/O 开销较大；而 Spark 在执行数据处理时，要将数据加载到内存中，直接在内存中加载中间结果数据集，减少了磁盘的 I/O 开销。

（4）数据容错。MapReduce 计算的中间结果数据，保存在磁盘中，Hadoop 底层实现了备份机制，从而保证了数据容错。Spark RDD 实现了基于 Lineage 的容错机制和设置检查点方式的容错机制，弥补数据在内存处理时因断电而导致数据丢失的缺陷。

7.2　Spark 的安装

7.2.1　Spark 的部署方式

（1）Standalone 模式。

Standalone 模式被称为集群单机模式，该模式下，Spark 集群架构为主从模式，即一台 Master 节点与多台 Slave 节点，Slave 节点启动的进程名称为 Worker，存在单点故障的问题。

（2）YARN 模式。

YARN 模式又被称为 Spark on YARN 模式，即把 Spark 作为一个客户端，将作业提交给 YARN 服务。由于在生产环境中，很多时候都要与 Hadoop 使用同一个集群，因此采用 YARN 来管理资源调度，可以提高资源利用率。

（3）Mesos 模式。

Mesos 是一款资源调度管理系统，类似于 YARN，提供了有效的、跨分布式应用或框架的资源隔离和共享服务。

7.2.2 Spark 的安装

安装 Spark 集群前,需要安装 Hadoop 环境(见第 2 章)。本书 Spark 的安装采用 Standalone 集群模式,采用如下配置环境:一台 Master 节点和两台 Slave 节点。其中,主机名 bigdata01 是 Master 节点,bigdata02 和 bidgdata03 是 Slave 节点。操作系统及其他安装版本如下。

Linux 系统:CentOS_ 7.0 版本。
Hadoop:2.7.7 版本。
JDK:1.8。
Scala:2.11.8。
Spark:2.1.0。
具体安装过程如下。

(1) 下载安装包,其中 Scala 和 Spark 下载地址如下:

https://downloads.lightbend.com/scala/2.11.8/scala-2.11.8.tgz
https://archive.apache.org/dist/spark/spark-2.1.0/spark-2.1.0-bin-hadoop2.7.tgz

(2) 解压安装包到指定目录。在 CentOS 命令行中输入如下命令:

```
tar -zxvf scala-2.11.8.tgz -C /hadoop/
tar -zxvf spark-2.1.0-bin-hadoop2.7.tgz -C /hadoop/
```

(3) 修改配置文件。

① 修改 /etc/profie 文件,增加如下内容:

```
export SCALA_HOME=/hadoop/scala-2.11.8
export PATH=$SCALA_HOME/bin:$PATH
export SPARK_HOME=/hadoop/spark-2.1.0-bin-hadoop2.7
export PATH=$PATH:$SPARK_HOME/bin
```

② 配置生效,在 CentOS 命令行中输入如下命令:

```
source /etc/profile
```

③ 修改 spark-env.sh 文件。

复制 spark-env.sh.template 成 spark-env.sh,在 CentOS 命令行中输入如下命令:

```
cp spark-env.sh.template spark-env.sh
```

④ 修改 $SPARK_HOME/conf/spark-env.sh 文件,添加如下内容:

```
export JAVA_HOME=/usr/java/jdk1.8.0_161
export SCALA_HOME=/hadoop/scala-2.11.8
export HADOOP_HOME=/hadoop/hadoop-2.7.7
export HADOOP_CONF_DIR=/hadoop/hadoop-2.7.7/etc/hadoop
export SPARK_MASTER_IP=172.16.106.69
export SPARK_MASTER_HOST=172.16.106.69
export SPARK_LOCAL_IP=172.16.106.69
```

```
export SPARK_WORKER_MEMORY = 1g
export SPARK_WORKER_CORES = 2
export SPARK_HOME = /hadoop/spark-2.1.0-bin-hadoop2.7
exportSPARK_LOCAL_DIRS = /hadoop/spark-2.1.0-bin-hadoop2.7
export SPARK_DIST_CLASSPATH = $(//hadoop/hadoop-2.7.7/bin/hadoop classpath)
```

⑤ 复制 slaves.template 文件，重新命名为 slaves，在 CentOS 命令行中输入如下命令：

```
cp slaves.template slaves
```

⑥ 修改 $SPARK_HOME/conf/slaves 文件，添加如下内容：

```
bigdata01
bigdata02
bigdata03
```

⑦ 将配置好的 Spark 安装文件复制到 bigdata02 和 bigdata03 节点，在 CentOS 命令行中输入如下命令：

```
scp -r /hadoop/scala-2.11.8/ root@bigdata02:/hadoop/
scp -r /hadoop/scala-2.11.8/ root@bigdata03:/hadoop/
scp -r /hadoop/spark-2.1.0-bin-hadoop2.7/ root@bigdata02:/hadoop/
scp -r /hadoop/spark-2.1.0-bin-hadoop2.7/ root@bigdata03:/hadoop/
```

⑧ 修改 bigdata02 和 bigdata03 配置。

在 bigdata02 和 bigdata03 上分别修改 /etc/profile 文件，增加 Spark 的配置。在 bigdata02 和 bigdata03 上修改 $SPARK_HOME/conf/spark-env.sh，将 export SPARK_LOCAL_IP 改成 bigdata02 和 bigdata03 对应节点的 IP。

（4）在 Master 节点启动集群，在 CentOS 命令行中输入如下命令：

```
/hadoop/spark-2.1.0-bin-hadoop2.7/sbin/start-all.sh
```

运行结果如下：

```
[root@bigdata01 sbin]# ./start-all.sh
starting org.apache.spark.deploy.master.Master, logging to /hadoop/spark-2.1.0-bin-hadoop2.7/logs/spark-root-org.apache.spark.deploy.master.Master-1-bigdata01.out
bigdata03: starting org.apache.spark.deploy.worker.Worker, logging to /hadoop/spark-2.1.0-bin-hadoop2.7/logs/spark-root-org.apache.spark.deploy.worker.Worker-1-bigdata03.out
bigdata02: starting org.apache.spark.deploy.worker.Worker, logging to /hadoop/spark-2.1.0-bin-hadoop2.7/logs/spark-root-org.apache.spark.deploy.worker.Worker-1-bigdata02.out
bigdata01: starting org.apache.spark.deploy.worker.Worker, logging to /hadoop/spark-2.1.0-bin-hadoop2.7/logs/spark-root-org.apache.spark.deploy.worker.Worker-1-bigdata01.out
```

（5）使用 jps 命令查看集群是否启动成功，运行结果如下：

```
[root@bigdata01 sbin]# jps
32240 DFSZKFailoverController
13680 Worker
13539 Master
31687 NameNode
32023 JournalNode
32503 NodeManager
31304 QuorumPeerMain
32376 ResourceManager
31805 DataNode
```

（6）在浏览器上输入 http://172.16.106.69:8080/，其中 172.16.106.69 为 Spark 的 Master 的 IP 地址，Spark 集群管理界面如图 7-2 所示。

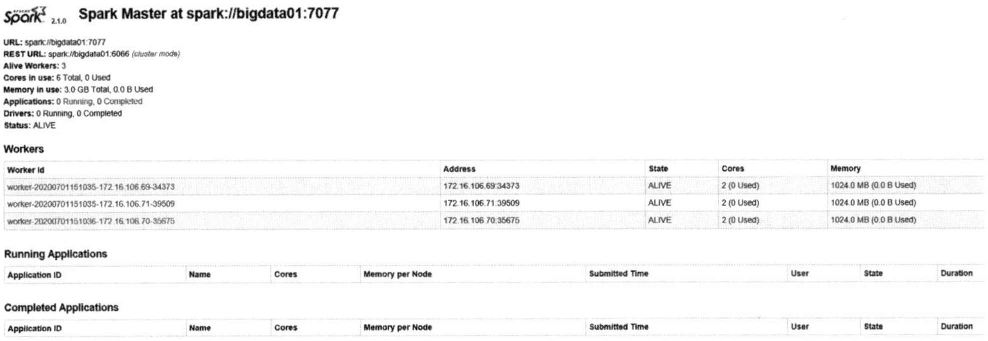

图 7-2 Spark 集群管理界面

7.3 Spark 运行架构与原理

Spark 运行架构主要由 SparkContext、Cluster Manager 和 Worker 组成。其中，Cluster Manager 负责整个集群的统一资源管理；Worker 节点中的 Executor 是应用执行的主要进程，内部含有多个 Task 线程以及内存空间。Spark 运行架构如图 7-3 所示。

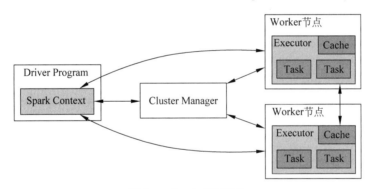

图 7-3 Spark 运行架构

（1）Program：用户编写的 Spark 应用程序。

（2）Driver：运行应用程序的 main() 函数并创建 SparkContext。

（3）SparkContext：准备 Spark 应用程序的运行环境，在 Spark 中由 SparkContext 负责与 Cluster Manager 通信，进行资源申请、任务的分配和监控等，当 Executor 部分运行完毕后，Driver 同时负责将 SparkContext 关闭。

（4）Executor：运行在工作节点（WorkerNode）的一个进程，负责运行 Task。

（5）Task：运行在 Executor 上的工作单元。

（6）Cluter Manager：在集群上获取资源的外部服务。目前有三种类型。

① Standalone：Spark 原生的资源管理，由 Master 负责资源的分配。

② Apache Mesos：与 Hadoop MapReduce 兼容性良好的一种资源调度框架。

③ Hadoop YARN：主要是指 YARN 中的 ResourceManager。

当执行一个 Application 时，Driver 会向集群管理器申请资源，启动 Executor，并向

Executor 发送应用程序代码和文件,然后在 Executor 上执行 Task,运行结束后,执行结果会返回给 Driver,或者写到 HDFS 或者其他数据库中。

与 Hadoop MapReduce 计算框架相比,Spark 所采用的 Executor 有如下两个优点。

(1)利用多线程来执行具体的任务,减少任务的启动开销。

(2)Executor 中有一个 BlockManager 存储模块,会将内存和磁盘共同作为存储设备,有效减少 I/O 开销。

7.4 Spark 运行流程

Spark 应用在集群上作为独立的进程组来运行,Spark 运行流程如图 7-4 所示。

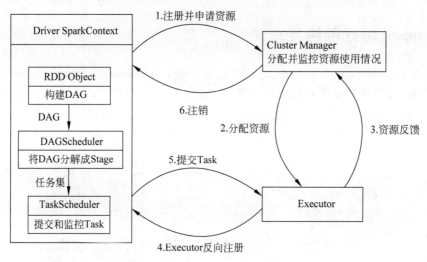

图 7-4　Spark 运行流程

(1)当一个 Spark 应用被提交时,根据提交参数创建 Driver 进程,为应用构建起基本的运行环境,即由 Driver 创建一个 SparkContext 进行资源的申请、任务的分配和监控。

(2)SparkContext 根据 RDD 的依赖关系构建 DAG,DAG 提交给 DAGScheduler 分解成 Stage,然后把一个个 Task 提交给底层调度器 TaskScheduler 处理。

(3)资源管理器 Cluster Manager 为 Executor 分配资源,并启动 Executor 进程。

(4)Executor 向 SparkContext 申请 Task,TaskScheduler 将 Task 发放给 Executor 运行并提供应用程序代码。

(5)Task 在 Executor 上运行把执行结果反馈给 TaskScheduler,然后反馈给 DAGScheduler,运行完毕后写入数据并释放所有资源。

7.5 本章小结

本章主要介绍了 Spark 的基本概念和主要特点,简单介绍了 Spark 安装、运行架构和运行流程,为学习 Spark RDD 和 Spark SQL 做基础知识储备。

第 8 章

Spark RDD 弹性分布式数据集

8.1 RDD 的设计与运行原理

8.1.1 RDD 的概念

RDD(Resilient Distributed Dataset,即弹性分布式数据集)是一个容错的、并行的数据结构,可以让用户显式地将数据存储到磁盘和内存中,并且还能控制数据的分区。不同 RDD 之间可以通过转换操作形成依赖关系实现管道化,从而避免了中间结果的 I/O 操作,提高数据处理的速度和性能。

一个 RDD 就是一个分布式对象集合,本质上是一个只读的分区记录集合。每个 RDD 可以分成多个分区,每个分区就是一个数据集片段,并且一个 RDD 的不同分区可以被保存到集群中不同的节点上,从而可以在集群中的不同节点上进行并行计算。RDD 提供了一种高度受限的共享内存模型,即 RDD 是只读的记录分区的集合,不能直接修改,只能基于稳定的物理存储中的数据集来创建 RDD,或者通过在其他 RDD 上执行确定的转换操作(如 Map、join 和 groupBy)而创建得到新的 RDD。RDD 提供了一组丰富的操作以支持常见的数据运算,分为"行动"(Action)和"转换"(Transformation)两种类型。前者用于执行计算并指定输出的形式;后者用于指定 RDD 之间的相互依赖关系。两类操作的主要区别是,转换操作(比如 Map、filter、groupBy、join 等)接收 RDD 并返回 RDD,而行动操作(比如 count、collect 等)接收 RDD 但是返回非 RDD(即输出一个值或结果)。RDD 提供的转换接口都非常简单,都是类似 Map、filter、groupBy、join 等粗粒度的数据转换操作,而不是针对某个数据项的细粒度修改。因此,RDD 比较适合对于数据集中元素执行相同操作的批处理式应用,而不适合用于需要异步、细粒度状态的应用,比如 Web 应用系统、增量式的网页爬虫等。正因为这样,这种粗粒度转换接口设计,会使人直觉上认为 RDD 的功能很受限、不够强大。但是，实际上 RDD 已经被实践证明可以很好地应用于许多并行计算应用中，可以具备很多现有计算框架的表达能力，并且可以应用于这些框架处理不了的交互式数据挖掘应用。

8.1.2　RDD 的分区

通常 RDD 很大，会被分成很多个分区，分别保存在不同的节点上。RDD 分区的一个原则是分区的个数尽量等于集群中的 CPU 内核数目。

对于不同的 Spark 部署模式而言，都可以通过设置 spark.default.parallelism 参数的值配置默认的分区数目。一般而言，本地模式，默认为本地机器的 CPU 数目，若设置了 local[N]，则默认为 N；Apache Mesos 模式，默认的分区数为 8；Standalone 或 YARN 模式，集群中所有 CPU 核心数目总和与 2 二者中取较大值作为默认值。

8.1.3　RDD 的依赖关系

1. 有向无环图

DAG(Directed Acyclic Graph)叫作有向无环图，Spark 中的 RDD 通过一系列的转换算子操作和行动算子操作形成了一个 DAG。DAG 是一种非常重要的图论数据结构。如果一个有向图无法从任意顶点出发经过若干条边回到该点，则这个图就是有向无环图。有向无环图示例如图 8-1 所示。"4→6→1→2"是一条路径，"4→6→5"也是一条路径，并且图中不存在从顶点经过若干条边后能回到该点。

2. 宽依赖与窄依赖

在 DAG 中，最初的 RDD 被称为基础 RDD，在基础 RDD 之上使用算子的过程中后续生成的 RDD 被称为一个子 RDD，它们之间存在依赖关系。无论哪个 RDD 出现问题，都可以由这种依赖关系重新计算而成。这种依赖关系就被称为 RDD 血统(Lineage)。血统的表现方式主要分为宽依赖与窄依赖。窄依赖示意图如图 8-2 所示。

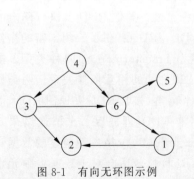

图 8-1　有向无环图示例　　　　图 8-2　窄依赖示意图

窄依赖是指父 RDD 的每一个分区最多被一个子 RDD 的分区使用。窄依赖的表现一般分为两类：第一类表现为一个父 RDD 的分区对应于一个子 RDD 的分区；第二类表现为多个父 RDD 的分区对应于一个子 RDD 的分区。

RDD 做 Map、filter 和 union 算子操作时，是属于窄依赖的第一类表现；而 RDD 做 join 算子操作（对输入进行协同划分）时，是属于窄依赖表现的第二类。输入协同划分是指多个父 RDD 的某一个分区的所有 Key，被划分到子 RDD 的同一分区。当子 RDD 做算子操作，因为某个分区操作失败导致数据丢失时，只需要重新对父 RDD 中对应的分区做算子操作即可恢复数据。

宽依赖是指子 RDD 的每一个分区都会使用所有父 RDD 的所有分区或多个分区。宽依赖示意图如图 8-3 所示。

父 RDD 做 groupByKey 和 join（输入未协同划分）算子操作时，子 RDD 的每一个分区都会依赖于所有父 RDD 的所有分区。当子 RDD 做算子操作，因为某个分区操作失败导致数据丢失时，则需要重新对父 RDD 中的所有分区进行算子操作才能恢复数据。

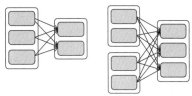

图 8-3　宽依赖示意图

3. Shuffle

窄依赖与宽依赖的区别是是否发生 Shuffle（洗牌）操作。宽依赖会发生 Shuffle 操作，而窄依赖不会发生 Shuffle 操作。在第 4 章介绍 MapReduce 框架时，知道 Shuffle 是连接 Map 和 Reduce 之间的桥梁，Map 的输出要用到 Reduce 中必须经过 Shuffle 这个环节，Shuffle 的性能高低直接影响了整个程序的性能和吞吐量。Spark 的 Shuffle 借鉴了 Hadoop 的 MapReduce 思想，但是 Spark 的 Shuffle 立足于 RDD，以 RDD 分区为单位进行迭代计算，可以通过血统图以及检查点（Checkpoint）实现高效容错。

Spark Shuffle 一般分为两个部分：Shuffle Write 和 Shuffle Fetch。前者是 Map 任务划分分区，输出中间结果；而后者则是 Reduce 任务获取到的这些中间结果。Spark 中 Shuffle 的流程如下：

（1）每一个 Mapper 会根据 Reducer 的数量创建出相应的 Bucket（桶），Bucket 的数量是 $M \times R$，其中 M 是 Map 的个数，R 是 Reduce 的个数。

（2）Mapper 产生的结果会根据设置的分区算法填充到每个 Bucket 中去。这里的分区算法是可以自定义的，当然默认的算法是根据 Key 哈希到不同的 Bucket 中去。

（3）当 Reducer 启动时，它会根据自己任务 id 和所依赖的 Mapper 的 id 从远端或是本地的 Block Manager 中取得相应的 Bucket 作为 Reducer 的输入进行处理。这里的 Bucket 是一个抽象概念，在实现中每个 Bucket 可以对应一个文件或者对应文件的一部分。

4. DAG 调度阶段

根据 RDD 之间依赖关系的不同可将 DAG 划分成不同的调度阶段(Stage)。对窄依赖来说,RDD 分区的转换处理是在一个线程中完成的,所以窄依赖会被 Spark 划分到同一个 Stage 中;而对宽依赖来说,由于有 Shuffle 存在,因此只能在父 RDD 处理完成后,下一个 Stage 才能开始接下来的计算,因此宽依赖是划分 Stage 的依据,当 RDD 进行转换操作,遇到宽依赖类型的转换操作时,就划为一个 Stage。

DAG 调度阶段划分,如图 8-4 所示。其中,A、C、E 是三个 RDD 的实例。

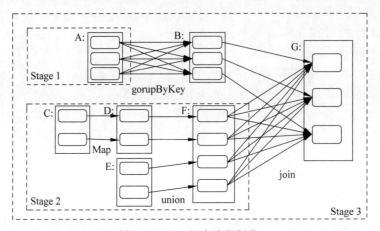

图 8-4 DAG 调度阶段划分

当 A 做 groupByKey 转换操作生成 B 时,由于 groupByKey 转换操作属于宽依赖类型,因此就把 A 划分为一个 Stage,即 Stage 1。

当 C 做 Map 转换操作生成 D 时,D 与 E 做 union 转换操作生成 F。由于 Map 和 union 转换操作都属于窄依赖类型,因此不进行 Stage 的划分,而是将 C、D、E、F 加入到同一个 Stage 中,即 Stage 2。当 F 与 B 进行 join 转换操作时,由于这时的 join 操作是非协同划分,因此属于宽依赖,所以会划分为一个 Stage,即 Stage 3。

8.1.4 RDD 在 Spark 中的运行流程

RDD 在 Spark 中的运行流程分为 RDD Objects、DAGScheduler、TaskScheduler 以及 Worker 4 个部分。

(1) RDD Object。当 RDD 对象创建后,SparkContext 会根据 RDD 对象构建 DAG,然后将 Task 提交给 DAGScheduler。

(2) DAGScheduler。将作业的 DAG 划分成不同 Stage,每个 Stage 都是 TaskSet 任务集合,并以 TaskSet 为单位提交给 TaskScheduler。

(3) TaskScheduler。通过 TaskSetManager 管理 Task,并通过集群中的资源管理器把 Task 发给集群中 Worker 的 Executor。

(4) Worker。Spark 集群中的 Worker 接收到 Task 后,把 Task 运行在 Executor 进程中,一个进程中可以有多个线程在工作,从而可以处理多个数据分区。

8.1.5 RDD 容错机制

RDD 容错主要采用如下两种方式。

(1) 血统方式。根据 RDD 之间的依赖关系对丢失数据的 RDD 进行数据恢复。若丢失数据的子 RDD 进行窄依赖运算,则只需要把丢失数据的父 RDD 的对应分区进行重新计算,不依赖其他节点,并且在计算过程中不存在冗余计算;若丢失数据的 RDD 进行宽依赖运算,则需要父 RDD 所有分区都进行从头到尾计算,计算过程中存在冗余计算。

(2) 检查点方式。其本质是将 RDD 写入磁盘存储。当 RDD 进行宽依赖运算时,只要在中间阶段设置一个检查点进行容错,即 Spark 中的 SparkContext 调用 setCheckpoint() 方法,设置容错文件系统目录作为检查点,将检查点的数据写入之前设置的容错文件系统中进行持久化存储。若后面有节点宕机导致分区数据丢失,则以从做检查点的 RDD 开始重新计算,不需要从头到尾的计算,从而减少开销。

8.2 RDD API 编程

Spark 用 Scala 语言实现了 RDD 的 API,开发人员可以通过调用 API 实现对 RDD 的各种操作。RDD 典型的执行过程如下。

(1) RDD 读入外部数据源(或者内存中的集合)进行创建。

(2) RDD 经过一系列的转换(Transformation)操作,每一次都会产生不同的 RDD,供给下一个转换使用。

(3) 最后一个 RDD 经行动(Action)操作进行处理,并输出到外部数据源(或者变成 Scala 集合或标量)。

需要说明的是,RDD 采用了惰性调用,即在 RDD 的执行过程中,真正的计算发生在 RDD 的行动操作,对于行动之前的所有转换操作,Spark 只是记录下转换操作应用的一些基础数据集以及 RDD 生成的轨迹,即相互之间的依赖关系,而不会触发真正的计算。

8.2.1 RDD 的创建

Spark 可以从 Hadoop 支持的任何存储源中加载数据去创建 RDD,包括本地文件系统和 HDFS 等文件系统。可以通过 Spark 中的 SparkContext 对象调用 textFile() 方法加载数据创建 RDD。首先启动 Hadoop,并在操作系统和 HDFS 上准备一个文本文件。准备完成后,运行 Spark-Shell。Spark-Shell 提供了一种学习 API 的简单方式,以及一个能够进行交互式分析数据的强大工具。Spark-Shell 运行命令及界面如图 8-5 所示。

Spark 支持从存储源中加载数据去创建 RDD,可以使用 Scala 或 Python 语言编写,主要有以下 3 种方式。

图 8-5　Spark-Shell 运行命令及界面

（1）从文件系统加载数据创建 RDD，Scala 语句及运行结果如下：

```
scala> val wordFile = sc.textFile("file:///words.txt");
wordFile: org.apache.spark.rdd.RDD[String] = file:///words.txt MapPartitionsRDD[1] at textFile at <console>:24
```

从运行结果反馈的信息可以看出，wordFile 是一个 String 类型的 RDD，或者以后可以简单称为 RDD[String]。也就是说，这个 RDD[String]里面的元素都是 String 类型。

（2）从 HDFS 上加载数据创建 RDD，Scala 语句及运行结果如下：

```
scala> val lines = sc.textFile("/input/word.txt")
lines: org.apache.spark.rdd.RDD[String] = /input/word.txt MapPartitionsRDD[10] at textFile at <console>:24
scala> val lines = sc.textFile("hdfs://localhost:9000/input/wc.txt")
lines: org.apache.spark.rdd.RDD[String] = hdfs://localhost:9000/input/wc.txt MapPartitionsRDD[12] at textFile at <console>:24
```

上面两种创建 RDD 的方式是完全等价的，只不过使用了不同的目录形式。

（3）通过并行集合创建 RDD。

Spark 可以通过并行集合创建 RDD。即从一个已经存在的集合、数组上，通过 SparkContext 对象调用 parallelize()方法创建 RDD。Scala 语句及运行结果如下：

```
scala> val array=Array(1,2,3,4,5,6)
array: Array[Int] = Array(1, 2, 3, 4, 5, 6)
scala> val arrRDD=sc.parallelize(array)
arrRDD: org.apache.spark.rdd.RDD[Int] = ParallelCollectionRDD[13] at parallelize at <console>:26
```

8.2.2　RDD 的操作

RDD 被创建好以后，在后续使用过程中一般会发生两种操作：一是转换，基于现有的数据集创建一个新的数据集；二是行动，在数据集上进行运算，返回计算值。

1. 转换

RDD 经过一系列的转换操作,每一次转换都会产生不同的 RDD,以供给下一次转换操作使用,直到最后一个 RDD 经过行动操作才会被真正计算处理,并输出外部数据源中,若是中间的数据结果需要复用,则可以进行缓存处理,将数据缓存到内存中。

RDD 处理过程中的转换操作主要用于根据已有 RDD 创建新的 RDD,每一次通过转换算子计算后都会返回一个新 RDD,供给下一个转换算子使用。RDD 转换算子操作常用 API 如表 8-1 所示。

表 8-1 RDD 转换算子操作常用 API

转换算子	相关说明
filter(func)	筛选出满足函数 func() 的元素,并返回一个新的数据集
Map(func)	将每个元素传递到函数 func() 中,返回的结果是一个新的数据集
flatMap(func)	与 Map() 相似,但是每个输入的元素都可以映射到 0 或者多个输出结果
groupByKey()	应用于<key,value>键值对的数据集时,返回一个新的<key,iterable<value>>形式的数据集
ReduceByKey(func)	应用于<key,value>键值对的数据集时,返回一个新的<key,value>形式的数据集。其中,每个 value 值是将每个 key 键传递到函数 func() 中进行聚合后的结果

2. 行动

行动算子主要将在数据集上运行计算后的数值返回到驱动程序,从而触发真正的计算。RDD 行动算子操作常用 API 如表 8-2 所示。

表 8-2 RDD 行动算子操作常用 API

转换算子	相关说明
count()	返回数据集中的元素个数
first()	返回数组的第一个元素
take(n)	以数组的形式返回数组集中的前 n 个元素
Reduce(func)	通过函数 func()(输入两个参数并返回一个值)聚合数据集中的元素
collect()	以数组的形式返回数据集中的所有元素
foreach(func)	将数据集中的每个元素都传递到函数 func() 中运行

3. RDD 操作示例:词频统计

在 Linux 系统的根目录下,有一个 words.txt 文件,文件中有多行文本,每行文本都是由 2 个单词构成的,且单词之间都是用空格分割。现在,通过 RDD 统计每个单词出现的次数(即词频)。词频统计操作过程如图 8-6 所示。

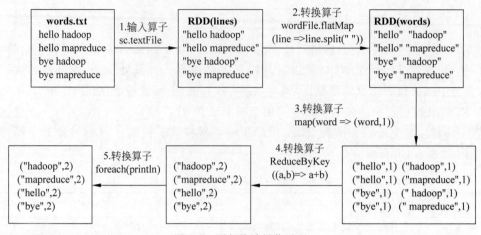

图 8-6　词频统计操作过程

8.3　程序示例：倒排索引

在本书的第 4 章用 MapReduce 框架实现了倒排索引，本例是用 Spark 的 RDD 操作实现倒排索引功能。具体操作步骤如下。

（1）利用 IntelliJ IDEA 新建一个 MAVEN 工程，ArtifactId 为 InvertedIndex，GroupId 为 com.sparklearn。

（2）修改 pom.xml 文件，添加相关依赖包，代码如下：

```xml
<?xml version = "1.0" encoding = "UTF-8"?>
<project xmlns = "http://maven.apache.org/POM/4.0.0"
    xmlns:xsi = "http://www.w3.org/2001/XMLSchema-instance"
    xsi:schemaLocation = "http://maven.apache.org/POM/4.0.0
    http://maven.apache.org/xsd/maven-4.0.0.xsd">
    <modelVersion>4.0.0</modelVersion>
    <groupId>com.spark</groupId>
    <artifactId>InvertedIndex</artifactId>
    <version>1.0-SNAPSHOT</version>
    <!-- 设置依赖版本号 -->
    <properties>
        <scala.version>2.11.8</scala.version>
        <hadoop.version>2.7.7</hadoop.version>
        <spark.version>2.1.0</spark.version>
    </properties>
    <dependencies>
        <!-- Scala -->
        <dependency>
            <groupId>org.scala-lang</groupId>
            <artifactId>scala-library</artifactId>
            <version>${scala.version}</version>
        </dependency>
```

```xml
<!-- Spark -->
<dependency>
    <groupId>org.apache.spark</groupId>
    <artifactId>spark-core_2.11</artifactId>
    <version>${spark.version}</version>
</dependency>
<!-- Hadoop -->
<dependency>
    <groupId>org.apache.hadoop</groupId>
    <artifactId>hadoop-client</artifactId>
    <version>${hadoop.version}</version>
</dependency>
<dependency>
    <groupId>org.apache.spark</groupId>
    <artifactId>spark-mllib_2.11</artifactId>
    <version>${spark.version}</version>
</dependency>
  </dependencies>
</project>
```

(3) 配置 Scala SDK，选择 File→Project Structure→Libraries 选项。配置 Scala SDK 界面如图 8-7 所示。

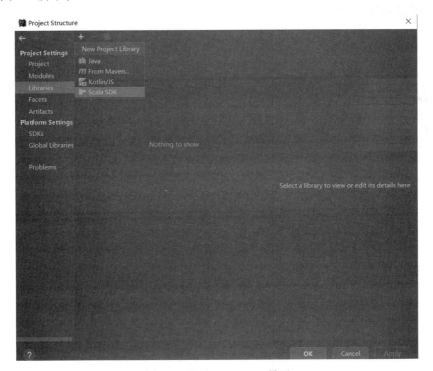

图 8-7　配置 Scala SDK 界面

(4) 新建名称为 scala 的文件夹，将文件夹设置成 Sources Root，设置 Sources Root 界面如图 8-8 所示。

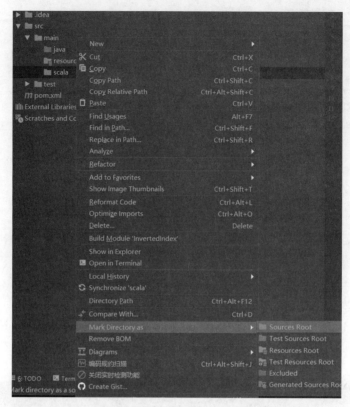

图 8-8 设置 Sources Root 界面

(5) 新建名为 InvertedIndex 的 Scala 类，主要功能是将 D 盘下的 file 文件夹下的三个文本文件中的文本按单词进行倒排索引。代码如下：

```
import org.apache.spark.sql.SparkSession
object InvertedIndex{
  def main(args: Array[String]): Unit = {
    //获取 SparkSession 对象
        val spark =
    SparkSession.builder().appName("InvertedIndex").master("local").getOrCreate()
    //读取目录
    val data = spark.sparkContext.wholeTextFiles("D:/file")
    val r1 = data.flatMap{
        x =>
            //使用分割"/"获取文件名
            val doc = x._1.split("/").last
            //先按行切分,再按列空格进行切分
        x._2.split("\r\n").flatMap(_.split(" ").Map { y => (y,doc)})}
    r1.foreach(println)
    //按单词分组
    val result = r1.groupByKey.Map{case(x,y) =>(x,y.toSet.mkString(":"))}
    result.foreach(println)
  }
}
```

InvertedIndex 程序运行结果如图 8-9 所示。

```
20/07/23 15:54:54 INFO Executor: Running task 0.0 in stage 3.0 (TID 3)
20/07/23 15:54:54 INFO ShuffleBlockFetcherIterator: Getting 1 non-empty blocks out of 1 blocks
20/07/23 15:54:54 INFO ShuffleBlockFetcherIterator: Started 0 remote fetches in 3 ms
(excellent,file1.txt)
(spark,file1.txt:file3.txt)
(is,file1.txt:file3.txt)
(hello,file2.txt)
(Mapreduce,file1.txt:file2.txt:file3.txt)
(simple,file1.txt)
(who,file3.txt)
(better,file1.txt:file3.txt)
(and,file3.txt)
20/07/23 15:54:54 INFO Executor: Finished task 0.0 in stage 3.0 (TID 3). 1632 bytes result sent to driver
```

图 8-9 InvertedIndex 程序运行结果

8.4 本章小结

本章主要介绍了 RDD 的设计与运行原理，并对 RDD 的基本操作进行了详细的介绍，最后用一个例子演示了用 Scala 编程实现对 RDD 的操作，以加深读者对本章的理解。

第 9 章

Spark SQL

9.1 Spark SQL 概述

9.1.1 Spark SQL 简介

Spark SQL 是 Spark 的一个模块，主要用于进行结构化数据的 SQL 查询引擎，开发人员能够通过使用 SQL 语句，实现对结构化数据的处理。如果开发人员不了解 Scala 语言和 Spark 常用 API，可以使用 Spark SQL 完成大数据分析。Spark SQL 的前身为 Shark，最初设计成与 Hive 兼容，但该项目于 2014 年因设计问题终止研发，转向 Spark SQL。Spark SQL 全面继承了 Shark，并进行了优化。

Spark SQL 主要提供了以下三个功能。

(1) Spark SQL 可从各种结构化数据源中读取数据，进行数据分析。

(2) Spark SQL 包含行业标准的 JDBC 和 ODBC 连接方式，因此它不局限于在 Spark 程序内使用 SQL 语句进行查询。

(3) Spark SQL 可以无缝地将 SQL 查询与 Spark 程序进行结合，它能够将结构化数据作为 Spark 中的分布式数据集（RDD）进行查询。

9.1.2 Spark SQL 的架构

Spark SQL 的架构与 Hive 的架构相比，把底层的 MapReduce 执行引擎更改为 Spark，还修改了 Catalyst 优化器（Spark SQL 快速的计算效率得益于 Catalyst 优化器）。从 HQL 被解析成语法抽象树，执行计划生成和优化的工作全部交给 Spark SQL 的 Catalyst 优化器进行负责和管理。Spark SQL 的架构如图 9-1 所示。

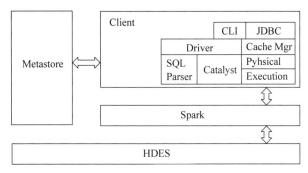

图 9-1　Spark SQL 的架构

9.2　DataFrame

9.2.1　DataFrame 简介

DataFrame 是一个以 RDD 为基础的分布式数据集，是 Spark SQL 使用的数据抽象。在 Spark 1.3.0 版本之前，DataFrame 被称为 SchemaRDD。DataFrame 使 Spark 具备处理大规模结构化数据的能力。DataFrame 的结构类似传统数据库的二维表格，可以从很多数据源中创建 DataFrame，如结构化文件、外部数据库、Hive 表等数据源。

DataFrame 可以看作是分布式的行对象的集合，在二维表数据集的每一列都带有名称和类型，这就是 Schema 元信息，这使得 Spark 框架可获取更多数据结构信息，从而对在 DataFrame 背后的数据源以及作用于 DataFrame 之上的数据变换进行有针对性的优化，最终达到提升计算效率的目的。

DataFrame 与 RDD 的区别如图 9-2 所示。从图 9-2 中可以看出 RDD 是分布式对象的集合，如 RDD[Person] 是以 Person 为类型参数，但是，Person 类的内部结构对于 RDD 而言却是不可知的。DataFrame 是一种以 RDD 为基础的分布式数据集，也就是分布式的 Row 对象的集合（每个 Row 对象代表一行记录），其提供了详细的结构信息，Spark SQL 可以清楚地知道该数据集中包含哪些列、每列的名称和类型。

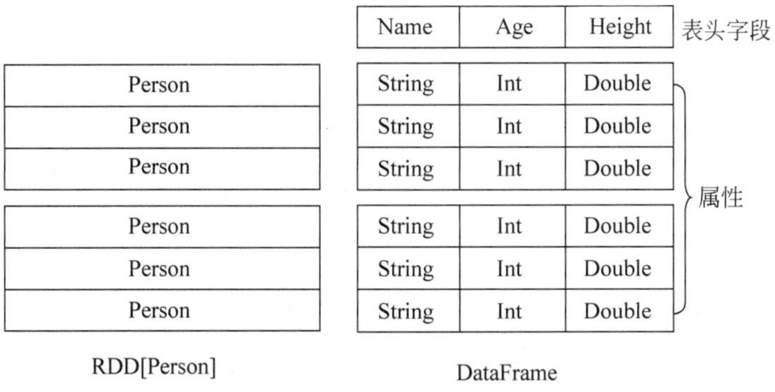

图 9-2　DataFrame 与 RDD 的区别

与 RDD 一样，DataFrame 的各种变换操作也采用惰性机制，只是记录了各种转换的逻辑转换路线图（是一个 DAG），不会发生真正的计算，这个 DAG 相当于一个逻辑查询计划，最终，会被翻译成物理查询计划，生成 RDD DAG，按照 RDD DAG 的执行方式去完成最终的计算得到结果。

9.2.2 DataFrame 的创建

创建 DataFrame 有以下两种基本方式。

（1）通过 Spark 读取数据源直接创建 DataFrame，使用 spark.read 从不同类型的文件中加载数据创建 DataFrame。spark.read 的具体操作如表 9-1 所示。

表 9-1 spark.read 的具体操作

方法名称	相关说明
spark.read.text("people.txt")	读取 TXT 格式文件，创建 DataFrame
spark.read.csv("people.csv")	读取 CSV 格式文件，创建 DataFrame
spark.read.json("people.json")	读取 JSON 格式文件，创建 DataFrame
spark.read.parquet("people.parquet")	读取 PARQUET 格式文件，创建 DataFrame

例如，在 Linux 操作系统 hadoop 文件夹下保存了两个文件，其中一个是 people.json，保存的是 JSON 格式的数据，内容如下：

{"name":"张三"}
{"name":"李四", "age":45}
{"name":"王五", "age":28}

另一个是 people.txt，保存的是普通文本格式数据，内容如下：

张三, 29
李四, 30
王五, 19.

利用 Spark Shell，读取 JSON 文件创建 DataFrame，示例代码及显示结果如下：

```
scala> val df = spark.read.json("file:///hadoop/people.json")
df: org.apache.spark.sql.DataFrame = [age: bigint, name: string]
scala> df.show();
+----+----+
| age|name|
+----+----+
|null|张三|
|  45|李四|
|  28|王五|
+----+----+
```

利用 Spark Shell，读取 TXT 文件创建 DataFrame，示例代码及显示结果如下：

```
scala> val dftxt = spark.read.json("file:///hadoop/people")
dftxt: org.apache.spark.sql.DataFrame = [_corrupt_record: string]
scala> dftxt.show();
+---------------+
|_corrupt_record|
+---------------+
|         张三, 29|
|         李四, 30|
|         王五, 19|
+---------------+
```

(2) 利用已存在的 RDD 调用 toDF() 方法转换得到 DataFrame,具体操作步骤如下。

① 从文件系统加载数据创建 RDD,代码如下:

```
val lineRDD = sc.textFile("file:///hadoop/people").Map(_.split(","))
```

② 定义类,代码如下:

```
case class Person(name:String,age:Int)
```

③ 实例化类并赋值,代码如下:

```
val personRDD = lineRDD.Map(x => Person(x(0), x(1).toInt))
```

④ 调用 toDF() 方法转换得到 DataFrame,代码如下:

```
val personDF = personRDD.toDF()
```

9.2.3 DataFrame 的常用操作

DataFrame 提供了两种语法风格:一种是领域特定语言(DSL)风格;另一种是 SQL 风格。DataFrame 的 DSL 风格操作的常用方法如表 9-2 所示。

表 9-2 DataFrame 的 DSL 风格操作的常用方法

方 法 名 称	相 关 说 明
show()	查看 DataFrame 中的具体内容信息
printSchema()	查看 DataFrame 的 Schema 信息
select()	查看 DataFrame 中选取部分列的数据及进行重命名
filter()	实现条件查询,过滤出想要的结果
groupBy()	对记录进行分组
sort()	对特定字段进行排序操作

(1) DSL 风格操作 DataFrame,示例代码及显示结果如下:

```
scala> val df = spark.read.json("file:///hadoop/people.json")
df: org.apache.spark.sql.DataFrame = [age: bigint, name: string]

scala> df.show()
+---+----+
|age|name|
+---+----+
| 18| 张三|
| 45| 李四|
| 28| 王五|
+---+----+

scala> df.printSchema()
root
 |-- age: long (nullable = true)
 |-- name: string (nullable = true)

scala> df.select(df("name"),df("age")+1).show()
+----+---------+
|name|(age + 1)|
+----+---------+
| 张三|       19|
| 李四|       46|
| 王五|       29|
+----+---------+
```

```
scala> df.filter(df("age") > 20 ).show()
+---+----+
|age|name|
+---+----+
| 45| 李四|
| 28| 王五|
+---+----+

scala> df.groupBy("age").count().show()
+---+-----+
|age|count|
+---+-----+
| 28|    1|
| 18|    1|
| 45|    1|
+---+-----+

scala> df.sort(df("age").desc).show()
+---+----+
|age|name|
+---+----+
| 45| 李四|
| 28| 王五|
| 18| 张三|
+---+----+
```

（2）SQL 风格操作 DataFrame。SQL 风格操作 DataFrame 需要先将 DataFrame 注册成一个临时表，然后使用 spark.sql 进行 SQL 风格操作，示例代码及显示结果如下：

```
scala> df.registerTempTable("t_person")
warning: there was one deprecation warning; re-run with -deprecation for details
scala> spark.sql("select * from t_person order by age desc limit 2").show()
+---+----+
|age|name|
+---+----+
| 45| 李四|
| 28| 王五|
+---+----+
```

9.3 Dataset

Dataset 是从 Spark 1.6 Alpha 版本中引入的一个新的数据抽象结构，最终在 Spark 2.0 版本被定义成 Spark 新特性。Dataset 提供了特定域对象中的强类型集合，也就是在 RDD 的每行数据中添加类型约束条件。Dataset 结合了 RDD 和 DataFrame 的优点，并且可以调用封装的方法以并行方式进行转换等操作。

Dataset 数据的表现形式主要有两种，其中一种是在 RDD 每行数据的基础之上，添加一个数据类型（value：String）作为 Schema 元数据信息，示例代码及显示结果如下：

```
scala> val personDs=spark.createDataset(sc.textFile("file:///hadoop/people"))
personDs: org.apache.spark.sql.Dataset[String] = [value: string]

scala> personDs.show();
+------+
| value|
+------+
|张三,29|
|李四,30|
|王五,19|
+------+
```

可以对 Dataset 每行数据添加强数据类型，示例代码及显示结果如下：

```
scala> val df = spark.read.json("file:///hadoop/people.json")
df: org.apache.spark.sql.DataFrame = [age: bigint, name: string]

scala> case class Person(name:String,age:Long)
defined class Person

scala> val ds = df.as[Person];
ds: org.apache.spark.sql.Dataset[Person] = [age: bigint, name: string]

scala> ds.show
+---+----+
|age|name|
+---+----+
| 18| 张三|
| 45| 李四|
| 28| 王五|
+---+----+
```

9.4 Spark SQL 编程

9.4.1 DataFrame 操作

（1）新建工程。利用 IntelliJ IDEA 新建一个 MAVEN 工程，工程名为 sparkSqldemo，GroupId 为 com.sparkstudy。

（2）添加依赖包。修改 pom.xml 文件，添加相关依赖包，代码如下：

```xml
<?xml version = "1.0" encoding = "UTF-8"?>
<project xmlns = "http://maven.apache.org/POM/4.0.0"
    xmlns:xsi = "http://www.w3.org/2001/XMLSchema-instance"
    xsi:schemaLocation = "http://maven.apache.org/POM/4.0.0
    http://maven.apache.org/xsd/maven-4.0.0.xsd">
    <modelVersion>4.0.0</modelVersion>
    <groupId>com.sparkstudy</groupId>
    <artifactId>sparkSqlDemo</artifactId>
    <version>1.0-SNAPSHOT</version>
    <!-- 设置依赖版本号 -->
    <properties>
        <scala.version>2.11.8</scala.version>
        <hadoop.version>2.7.7</hadoop.version>
        <spark.version>2.1.0</spark.version>
    </properties>
    <dependencies>
        <!-- Scala -->
        <dependency>
            <groupId>org.scala-lang</groupId>
            <artifactId>scala-library</artifactId>
            <version>${scala.version}</version>
        </dependency>
        <!-- Spark -->
        <dependency>
            <groupId>org.apache.spark</groupId>
            <artifactId>spark-core_2.11</artifactId>
            <version>${spark.version}</version>
        </dependency>
        <!-- Hadoop -->
        <dependency>
```

```
            <groupId>org.apache.hadoop</groupId>
            <artifactId>hadoop-client</artifactId>
            <version>${hadoop.version}</version>
        </dependency>
        <dependency>
            <groupId>org.apache.spark</groupId>
            <artifactId>spark-mllib_2.11</artifactId>
            <version>${spark.version}</version>
        </dependency>
    </dependencies>
</project>
```

（3）配置 Scala SDK。选择 File→Project Structure→Libraries 选项。配置 Scala SDK 界面如图 9-3 所示。

图 9-3　配置 Scala SDK 界面

（4）新建文件夹。新建名称为 scala 的文件夹，将文件夹设置成 Sources Root，设置 Sources Root 界面如图 9-4 所示。

（5）新建类。新建名为 sparkSqlSchema 的 Scala 类，主要功能是读取 D 盘下的 people.txt 文件，使用编程方式操作 DataFrame，相关代码及说明如下：

```
import org.apache.spark.rdd.RDD
import org.apache.spark.sql.{DataFrame, SparkSession}
case class Person(name:String,age:Long)
```

第9章 Spark SQL

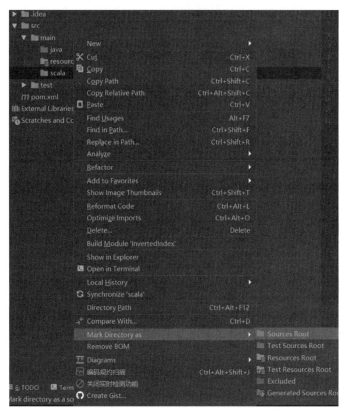

图 9-4 设置 Sources Root 界面

```
object sparkSqlSchema{
  def main(args: Array[String]): Unit = {
    //创建 Spark 运行环境
    val spark = SparkSession.builder().appName("sparkSqlSchema").master("local").getOrCreate()
    val sc = spark.sparkContext;
    //读取文件
    val data: RDD[Array[String]] = sc.textFile("D:/people.txt").Map(x => x.split(","));
    //将 RDD 与样例类关联
    val personRdd: RDD[Person] = data.Map(x => Person(x(0),x(1).toLong))
    //手动导入隐式转换
    import spark.implicits._
    val personDF: DataFrame = personRdd.toDF
    //显示 DataFrame 的数据
    personDF.show()
    //显示 DataFrame 的 schema 信息
    personDF.printSchema()
    //显示 DataFrame 记录数
    println(personDF.count())
    //显示 DataFrame 的所有字段
    personDF.columns.foreach(println)
    //取出 DataFrame 的第一行记录
```

```
            println(personDF.head())
            //显示 DataFrame 中 name 字段的所有值
            personDF.select("name").show()
            //过滤出 DataFrame 中年龄大于 20 的记录
            personDF.filter($"age" > 20).show()
            //统计 DataFrame 中年龄大于 20 的人数
            println(personDF.filter($"age" > 20).count())
            //统计 DataFrame 中按照年龄进行分组,求每个组的人数
            personDF.groupBy("age").count().show()
            //将 DataFrame 注册成临时表
            personDF.createOrReplaceTempView("t_person")
            //传入 SQL 语句,进行操作
            spark.sql("select * from t_person").show()
            spark.sql("select * from t_person where name = '王五'").show()
            spark.sql("select * from t_person order by age desc").show()
            //DataFrame 转换为 Dataset
            var ds = personDF.as[Person]
            ds.show()
            //关闭操作
            sc.stop()
            spark.stop()
        }
    }
```

sparkSqlSchema 程序运行结果如图 9-5 所示。

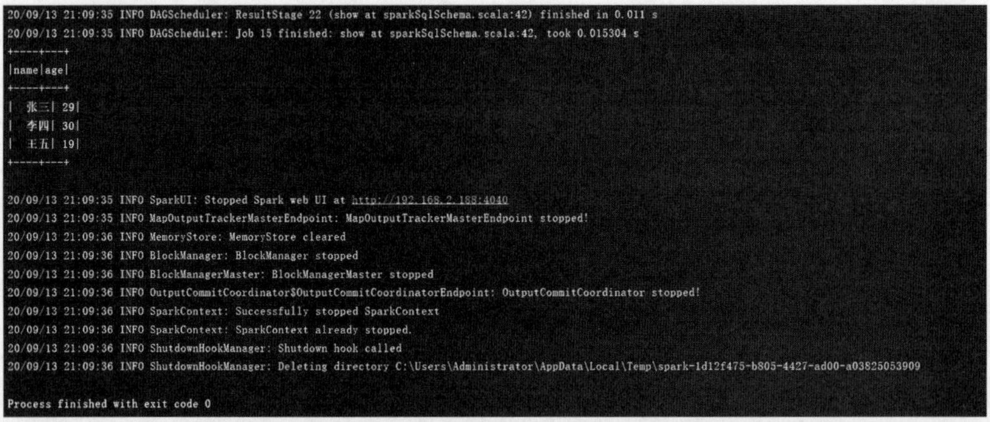

图 9-5 sparkSqlSchema 程序运行结果

9.4.2　Spark SQL 读写 MySQL 数据库

本示例程序使用 JDBC 连接 MySQL 数据库,并进行读写操作,主要步骤如下。

(1) 新建数据库。利用 Navicat 新建名为 spark 的数据库。新建数据库操作界面如图 9-6 所示。

(2) 新建表。在 spark 数据库下新建表 person,新建表操作界面如图 9-7 所示。

图 9-6 新建数据库操作界面

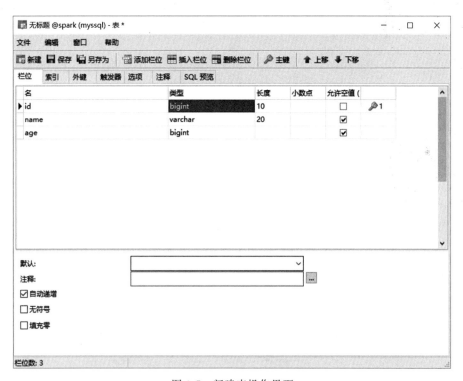

图 9-7 新建表操作界面

（3）添加依赖包。在 pom.xml 文件中添加连接 MySQL 数据库的依赖包，内容如下：

< dependency >
    < groupId > mysql </groupId >

```xml
        <artifactId>mysql-connector-java</artifactId>
        <version>5.1.38</version>
</dependency>
```

（4）新建类。新建 Scala 类 sparkSqlMysql，具体代码及说明如下：

```scala
import java.util.Properties
import org.apache.spark.rdd.RDD
import org.apache.spark.sql.{DataFrame,SaveMode,SparkSession}
case class Person(name:String,age:Long)
object sparkSqlMysql{
  def main(args: Array[String]): Unit = {
    //创建 SparkSession 对象
    val spark: SparkSession = SparkSession.builder()
      .appName("sparkSqlMysql")
      .master("local")
      .getOrCreate()
    val sc = spark.sparkContext
    //读取数据
    val data: RDD[Array[String]] = sc.textFile("D:/people.txt").Map(x => x.split(","));
    //RDD 关联 Person
    val personRdd: RDD[Person] = data.Map(x => Person(x(0), x(1).toLong))
    //导入隐式转换
    import spark.implicits._
    //将 RDD 转换为 DataFrame
    val personDF: DataFrame = personRdd.toDF()
    personDF.show()
    //创建 Properties 对象,配置连接 MySQL 的用户名和密码
    val prop = new Properties()
    prop.setProperty("user","root")
    prop.setProperty("password","123456")
    //将 personDF 写入 MySQL
    personDF.write.mode(SaveMode.Append).jdbc("jdbc:mysql://127.0.0.1:3306/spark?useUnicode=true&characterEncoding=utf8","person",prop)
    //从数据库中读取数据
    val mysqlDF: DataFrame = spark.read.jdbc("jdbc:mysql://127.0.0.1:3306/spark","person",prop)
    mysqlDF.show()
    spark.stop()
  }
}
```

（5）查看运行结果。运行 sparkSqlMysql 类，sparkSqlMysql 运行结果如图 9-8 所示。
查看 MySQL 数据库，可看到 person 表中添加了三条记录。parkSqlMysql 数据库运行结果如图 9-9 所示。

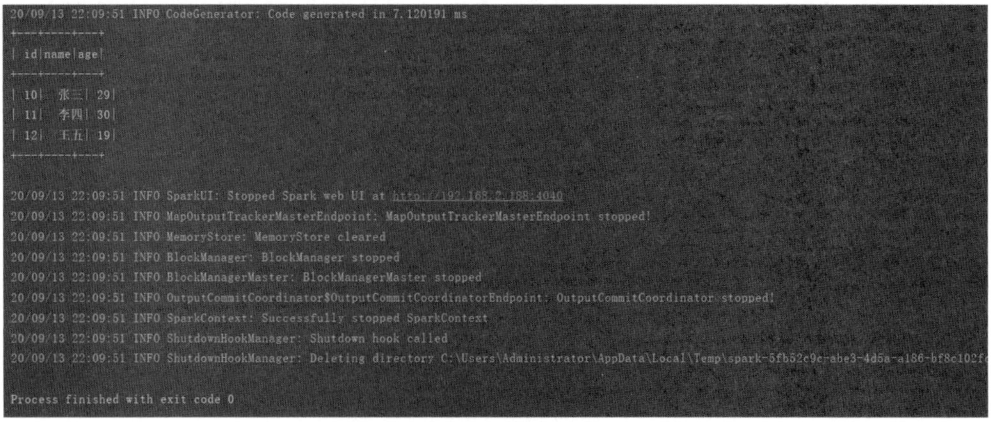

图 9-8 sparkSqlMysql 运行结果

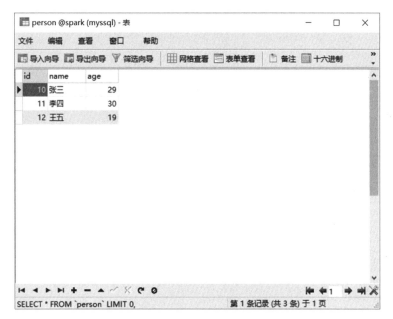

图 9-9 sparkSqlMysql 数据库运行结果

9.4.3 Spark SQL 读写 Hive

本示例程序连接 Hive，并读写 Hive 下的表，主要步骤如下。

（1）添加 Hive 依赖包。修改 pom.xml 文件，添加内容如下：

< dependency >
 < groupId > org.apache.spark </groupId >
 < artifactId > spark - hive_2.11 </artifactId >
 < version >${spark.version}</version >
</dependency >

(2) 连接 Hive。将安装 Hadoop 配置的 hdfs-site.xml 文件和安装 Hive 时配置的 hive-site.xml 文件复制到此工程的 resource 下。hive-site.xml 是用于让程序连接 Hive 的 MySQL 数据库,读取 Hive 的元数据信息,hdfs-site.xml 是为了让程序读到 Hadoop 的配置信息来访问 Hadoop。如果 hive-site.xml 中配置的 MySQL 数据库地址为 localhost,应将它配置成实际 IP 或主机名,如果配置成主机名,还需要修改 hosts 文件中加入 IP 和主机名的映射关系。hive-site.xml 配置如下:

```xml
<?xml version = "1.0" encoding = "UTF-8"?>
<configuration>
    <property>
        <name>javax.jdo.option.ConnectionURL</name>
        <value>jdbc:mysql://bigdata02:3306/metastore?createDatabaseIfNotExist = true</value>
        <description>JDBC connect string for a JDBC metastore</description>
    </property>
    <property>
        <name>javax.jdo.option.ConnectionDriverName</name>
        <value>com.mysql.jdbc.Driver</value>
        <description>Driver class name for a JDBC metastore</description>
    </property>
    <property>
        <name>javax.jdo.option.ConnectionUserName</name>
        <value>hadoop</value>
        <description>username to use against metastore database</description>
    </property>
    <property>
        <name>javax.jdo.option.ConnectionPassword</name>
        <value>hadoop</value>
        <description>password to use against metastore database</description>
    </property>
</configuration>
```

(3) 新建表。在 Hive 的 bigdata 数据库下新建 person 表,新建表语句及显示结果如下:

```
hive> use bigdata;
OK
Time taken: 0.012 seconds
hive> CREATE TABLE IF NOT EXISTS person (name string, age int)
    > row format delimited fields terminated by ',';
OK
Time taken: 0.051 seconds
```

(4) 新建 Scala 类 sparksqlToHIVE,其主要功能是读取 D 盘下的 people.txt 文件,使用编程方式操作 DataFrame,然后插入到 Hive 的 person 表中。相关代码及说明如下:

```scala
import org.apache.spark.rdd.RDD
import org.apache.spark.sql.{DataFrame, SparkSession}
case class Person(name: String, age: Long)
object sparksqlToHIVE{
  def main(args: Array[String]): Unit = {
    //设置访问用户名,主要用于访问 HDFS 下的 Hive warehouse 目录
    System.setProperty("HADOOP_USER_NAME", "root")
```

```scala
//创建 sparkSession
val spark: SparkSession = SparkSession.builder()
  .appName("sparksqlToHIVE")
  .config("executor-cores",1)
  .master("local")
  .enableHiveSupport() //开启支持 Hive
  .getOrCreate()
val sc = spark.sparkContext
//读取文件
  val data: RDD[Array[String]] = sc.textFile("D:/people.txt").Map(x => x.split(","));
//将 RDD 与样例类关联
val personRdd: RDD[Person] = data.Map(x => Person(x(0),x(1).toLong))
//手动导入隐式转换
import spark.implicits._
val personDF: DataFrame = personRdd.toDF
//显示 DataFrame 的数据
personDF.show()
//将 DataFrame 注册成临时表 t_person
personDF.createOrReplaceTempView("t_person")
//显示临时表 t_person 的数据
spark.sql("select * from t_person").show()
//使用 Hive 中 bigdata 数据库
spark.sql("use bigdata")
//将临时表 t_person 的数据插入使用 Hive 中 bigdata 数据库下的 person 表中
spark.sql("insert into person select * from t_person")
//显示用 Hive 中 bigdata 数据库下的 person 表数据
spark.sql("select * from person").show()
spark.stop()
  }
}
```

（5）查看运行结果。运行 sparksqlToHIVE 类，sparksqlToHIVE 类运行结果如图 9-10 所示。

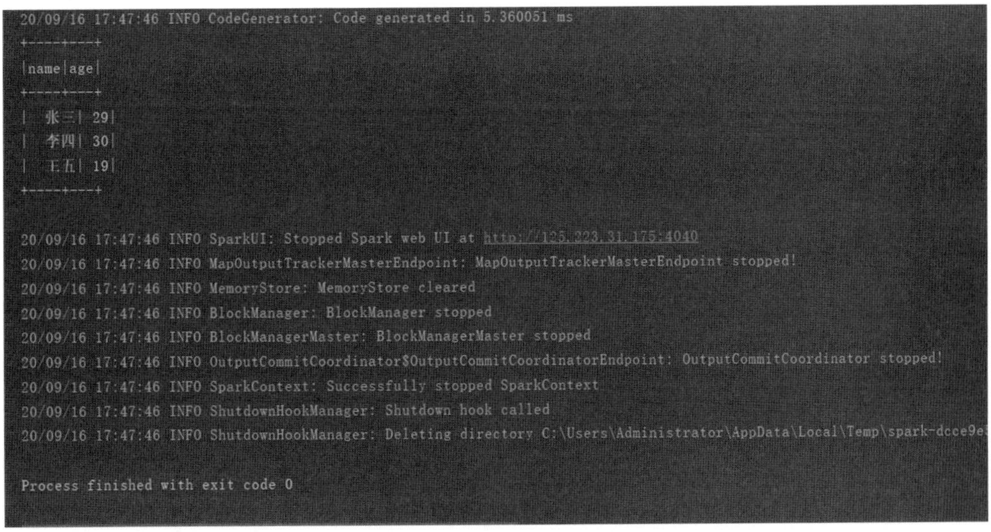

图 9-10　sparksqlToHIVE 类运行结果

9.5 本章小结

本章主要介绍了 Spark SQL 的原理和运行流程,并对 DataFrame 的基本操作进行了详细的介绍,最后通过三个例子演示了用 Scala 编程实现对 DataFrame 的操作,以加深读者对本章的理解。

第 10 章

Spark Streaming实时计算框架

10.1 Spark Streaming 概述

10.1.1 流数据和流计算

在大数据时代,数据可以分为静态数据和流数据。静态数据是指在很长一段时间内不会变化,一般不随运行而变化的数据。流数据是一组顺序、大量、快速、连续到达的数据序列,一般情况下数据流可被视为一个随时间延续而无限增长的动态数据集合。它应用于网络监控、传感器网络、航空航天、气象测控和金融服务等领域。对静态数据和流数据的处理,对应着两种截然不同的计算模式:批量计算和实时计算。批量计算以静态数据为对象,可以在很充裕的时间内对海量数据进行批量处理,计算得到有价值的信息。Hadoop就是典型的批处理模型,由 HDFS 和 HBase 存放大量的静态数据,由MapReduce 负责对海量数据执行批量计算。流数据必须采用实时计算。实时计算最重要的一个需求是能够实时得到计算结果,一般要求响应时间为秒级。当只需要处理少量数据时,实时计算并不是问题。但是,在大数据时代,不仅数据格式复杂、来源众多,而且数据量巨大,这就对实时计算提出了很大的挑战。因此,针对流数据的实时计算——流计算就应运而生。

一般来说,流计算秉承一个基本理念,即数据的价值随着时间的流逝而降低。因此,当事件出现时就应该立即进行处理,而不是缓存起来进行批量处理。为了及时处理流数据,就需要一个低延迟、可扩展、高可靠的处理引擎。对于一个流计算系统来说,它应达到如下需求。

(1) 高性能。处理大数据的基本要求,如每秒处理几十万条数据。

(2) 海量式。支持太字节(TB)级甚至是皮字节(PB)级的数据规模。

（3）实时性。必须保证一个较低的延迟时间，达到秒级别，甚至是毫秒级别。

（4）分布式。支持大数据的基本架构，能够平滑扩展。

（5）易用性。能够快速进行开发和部署。

（6）可靠性。能可靠地处理流数据。

目前，市场上有很多流计算框架，如 Twitter Storm 和 Yahoo! S4 等。Twitter Storm 是免费、开源的分布式实时计算系统，可简单、高效、可靠地处理大量的流数据；Yahoo! S4 开源流计算平台是通用的、分布式的、可扩展的、分区容错的、可插拔的流式系统。

流计算处理过程包括数据实时采集、数据实时计算和实时查询服务。

（1）数据实时采集。数据实时采集阶段通常采集多个数据源的海量数据，需要保证实时性、低延迟与稳定可靠。以日志数据为例，由于分布式集群的广泛应用，数据分散存储在不同的机器上，因此需要实时汇总来自不同机器上的日志数据。目前，有许多互联网公司发布的开源分布式日志采集系统均可满足每秒数百 MB 的数据采集和传输需求，如 Facebook 的 Scribe、LinkedIn 的 Kafka、淘宝的 TimeTunnel，以及基于 Hadoop 的 Chukwa 和 Flume 等。

（2）数据实时计算。流处理系统接收数据采集系统不断发来的实时数据，实时地进行分析计算，并反馈实时结果。

（3）实时查询服务。经由流计算框架得出的结果可供用户进行实时查询、展示或存储。流计算处理过程如图 10-1 所示。

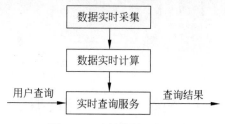

图 10-1　流计算处理过程

10.1.2　Spark Streaming 简介

Spark Streaming 是构建在 Spark 上的实时计算框架，且是对 Spark Core API 的一个扩展。它能够实现对流数据进行实时处理，并具有很好的可扩展性、高吞吐量和容错性。Spark Streaming 具有易用性、容错性及易整合性等显著特点。Spark Streaming 可整合多种输入数据源，如 Kafka、Flume、HDFS，甚至是普通的 TCP 套接字。经处理后的数据可存储至文件系统、数据库，或显示在仪表盘中。Spark Streaming 支持的输入输出数据源如图 10-2 所示。

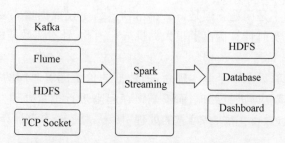

图 10-2　Spark Streaming 支持的输入输出数据源

10.1.3 DStream 简介

Spark Streaming 提供了一个高级抽象的流,即 DStream(离散流)。DStream 表示连续的数据流,可以通过 Kafka、Flume 等数据源创建,也可以通过现有 DStream 的高级操作创建。DStream 的内部是由一系列连续的 RDD 组成,每个 RDD 都是一小段时间分割开来的数据集。对 DStream 的任何操作,最终都会转变成对底层 RDD 的操作。例如,单词统计流式计算过程如图 10-3 所示。单词统计流式计算过程展示了进行单词统计时,每个时间片的数据(存储句子的 RDD)经 flatMap 操作,生成了存储单词的 RDD。整个流式计算可根据业务的需求对这些中间的结果进一步处理。单词统计流式计算过程如图 10-3 所示。

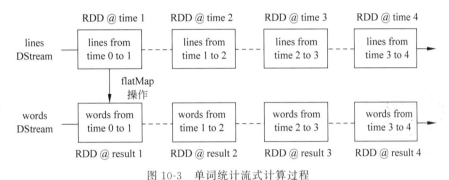

图 10-3 单词统计流式计算过程

10.2 DStream 编程

批处理引擎 Spark Core 把输入的数据按照一定的时间片(如 1s)分成一段一段的数据,每一段数据都会转换为 RDD 输入到 Spark Core 中,然后将 DStream 操作转换为 RDD 算子的相关操作,即转换操作以及输出操作。RDD 算子操作产生的中间结果数据会保存在内存中,也可以将中间的结果数据输出到外部存储系统中进行保存。

10.2.1 DStream 转换操作

DStream 转换操作可以分为无状态(Stateless)和有状态(Stateful)两种。

1. 无状态转换操作

无状态转换操作每个批次的处理不依赖于之前批次的数据。常见的 RDD 转化操作,例如 Map()、filter()、ReduceByKey()等,都是无状态转化操作。

需要注意的是,尽管这些函数看起来像作用在整个流上一样,但事实上每个 DStream 在内部是由许多 RDD 组成的,且无状态转换操作是分别应用到每个 RDD 上的。例如,ReduceByKey()会归约每个时间区间中的数据,但不会归约不同区间之间的数据。

无状态转换操作也能在多个 DStream 间整合数据,不过也是在各个时间区间内。例如,DStream 拥有和 RDD 一样的与连接相关的转换操作,也就是 cogroup()、join()等。

可以在 DStream 上使用这些操作,这样就对每个批次都分别执行了对应的 RDD 操作。也可以像在常规的 Spark 中一样使用 DStream 的 union() 操作将它和另一个 DStream 的内容合并起来。无状态转换操作常用方法如表 10-1 所示。

表 10-1 无状态转换操作常用方法

方法名称	相关说明
Map(func)	将源 DStream 的每个元素,传递到函数 func() 中进行转换操作,得到一个新的 DStream
flatMap(func)	与 Map() 相似,但是每个输入的元素都可以映射 0 或者多个输出结果
filter(func)	返回一个新的 DStream,仅包含源 DStream 中经过 func() 函数计算结果为 true 的元素
repartition(Partitions)	用于指定 DStream 分区的数量
union(other Stream)	返回一个新的 DStream,包含源 DStream 和其他 DStream 中的所有元素
count()	统计源 DStream 中每个 RDD 包含的元素个数,返回一个新 DStream
Reduce(func)	使用函数 func() 将源 DStream 中每个 RDD 的元素进行聚合操作,返回一个新 DStream
countByValue()	计算 DStream 中每个 RDD 内的元素出现的频次,并返回一个新的 DStream[(K,Long)]。其中,K 是 RDD 中元素的类型,Long 是元素出现的频次
join(otherStream,[numTasks])	当被调用类型分别为< key,value1 >和< key,value2 >键值对的两个 DStream 时,返回类型为< key,< value1,value2 >>键值对的一个新 DStream
cogroup(otherStream,[numTasks])	当被调用的两个 DStream 分别含有< key,value1 >和< key,value2 >键值对时,返回一个新 DStream
transform(func)	对源 DStream 中每个 RDD 应用 RDD-To-RDD() 函数返回一个新 DStream,在 DStream 中做任意 RDD 操作

2. 有状态转换操作

有状态转换操作需要使用 DStream 之前批次的数据或者是中间结果来计算当前批次的数据。有状态转换操作包括基于滑动窗口的转换操作和追踪状态变化的转换操作。有状态转换操作常用方法如表 10-2 所示。

表 10-2 有状态转换操作常用方法

方法名称	相关说明
updateStateByKey(func)	返回一个新状态 DStream,通过在键的先前状态和键的新值上应用给定函数 func() 更新每一个键的状态。该操作方法被用于维护每一个键的任意状态数据
window(windowLength,slideInterval)	返回基于源 DStream 的窗口进行批量计算后的一个新 DStream
countByWindow(windowLength,slideInterval)	返回基于滑动窗口的 DStream 中的元素数
ReduceByWindow(func,windowLength,slideInterval)	基于滑动窗口的源 DStream 中的元素进行聚合操作,返回一个新 DStream

续表

方法名称	相关说明
ReduceByKeyAndWindow(func,windowLength, slideInterval,[numTasks])	基于滑动窗口对<key,value>类型的 DStream 中的值,按 K 应用聚合函数 func()进行聚合操作,返回一个新 DStream
ReduceByKeyAndWindow (func, invFuncwindowLength,slideInterval,[numTasks])	更高效的 ReduceByKeyAndWindow()实现版本。每个窗口的聚合值,都是基于先前窗口的新数据进行聚合操作,并对离开窗口历史数据进行逆向聚合操作
countByValueAndWindow (windowLength, slideInterval,[numTasks])	基于滑动窗口计算源 DStream 中每个 RDD 内每个元素出现的频次,返回一个由<key,value>组成的新的 DStream

10.2.2 DStream 输出操作相关的方法

DStream 输出操作是将 DStream 中的内容输出到文件或其他外部系统中。输出操作常用方法如表 10-3 所示。

表 10-3 输出操作常用方法

方法名称	相关说明
saveAsHadoopFiles(prefix,[suffix])	将 DStream 中的内容以文本的形式保存为 Hadoop 文件,其中每次批处理间隔内产生的文件都以 prefix-TIME_IN_MS[.suffix]的方式命名
foreachRDD(func)	最基本的输出操作,将 func()函数应用于 DStream 中的 RDD 上,这个操作会输出数据到外部系统

10.3 DStream 编程示例

10.3.1 DStream 编程基本步骤——文件流

编写 DStream 程序主要有以下 5 个步骤。
(1) 通过创建输入 DStream 来定义输入源。
(2) 通过对 DStream 应用转换操作和输出操作来定义流计算。
(3) 用 streamingContext.start()来开始接收数据和处理流程。
(4) 通过 streamingContext.awaitTermination()方法来等待处理结束。
(5) 通过 streamingContext.stop()来手动结束流计算进程。

通过监视日志并对文件流进行操作,计算每个单词出现的次数的示例程序,演示编写 DStream 程序的基本步骤。具体过程如下。

(1) 新建 MAVEN 项目。其名称为 sparkDStream,新建过程详见第 9 章。在 pom.xml 文件中加入依赖包,具体代码如下:

```
<?xml version = "1.0" encoding = "UTF - 8"?>
<project xmlns = "http://maven.apache.org/POM/4.0.0"
```

```xml
xmlns:xsi = "http://www.w3.org/2001/XMLSchema-instance"
xsi:schemaLocation = "http://maven.apache.org/POM/4.0.0
http://maven.apache.org/xsd/maven-4.0.0.xsd">
    <modelVersion>4.0.0</modelVersion>
    <groupId>com.sparklearn</groupId>
    <artifactId>sparkDStream</artifactId>
    <version>1.0-SNAPSHOT</version>
    <properties>
        <spark.version>2.1.0</spark.version>
        <scala.version>2.11.8</scala.version>
    </properties>
    <dependencies>
        <dependency>
            <groupId>org.apache.spark</groupId>
            <artifactId>spark-core_2.11</artifactId>
            <version>${spark.version}</version>
        </dependency>
        <dependency>
            <groupId>org.apache.spark</groupId>
            <artifactId>spark-streaming_2.11</artifactId>
            <version>${spark.version}</version>
        </dependency>
    </dependencies>
</project>
```

(2) 新建 Scala 类。其名称为 WordCountStreaming，相应代码和说明如下：

```scala
import org.apache.spark._
import org.apache.spark.streaming._
object WordCountStreaming{
  def main(args: Array[String]){
    //设置为本地运行模式,有两个线程,其中一个监听,另一个处理数据
    val sparkConf = new SparkConf().setAppName("WordCountStreaming").setMaster("local[2]")
    //时间间隔为 20s
    val stc = new StreamingContext(sparkConf, Seconds(20))
    //定义输入源,监听本地目录,也可以采用 HDFS 文件
    val lines = stc.textFileStream("E:/log")
    //应用转换操作 flatMap 流计算
    val words = lines.flatMap(_.split(" "))
    //应用转换操作 Map()和 ReduceByKey()计算
    val wordCounts = words.Map(x => (x,1)).ReduceByKey(_ + _)
    wordCounts.print()
    //开始接收数据和处理流程
    stc.start()
    //等待处理结束
    stc.awaitTermination()
  }
}
```

(3) 运行程序验证结果。Spark Streaming 每隔 20s 就监听一次 E:/log 目录下在程

序启动后新增的文件,因为 Spark Streaming 不会去处理历史上已经存在的文件,所以为了能够让程序读取文件内容并显示到控制台,需要到 E:/log 目录下再新建一个 log3.txt 文件,在里面随便输入一些英文单词,本例输入的为:

```
hello hadoop
hello spark
spark bye
byehadoop
```

创建完成以后,切换到程序运行控制台。WordCountStreaming 运行结果如图 10-4 所示。

```
20/09/23 18:36:00 INFO DAGScheduler: Job 5 finished: print at WordCountStreaming.scala:15, took 0.023245 s
-------------------------------------------
Time: 1600857360000 ms
-------------------------------------------
(hello,2)
(bye,2)
(spark,2)
(hadoop,2)
```

图 10-4　WordCountStreaming 运行结果

10.3.2　无状态转换操作

1. cogroup 和 join 算子转换操作

cogroup 和 join 算子需要两个并行数据流,对两个数据流直接关联。不同的是 join 算子是把两个 RDD 按照相同的 key 拼在一起,类似 SQL 中的等值连接,可以类似地使用 leftOuterJoin、rightOuterJoin、fullOuterJoin 算子进行两个 RDD 的左连接、右连接和全连接。而 cogroup 算子是把两个 RDD 按照 key 拼起来,但是它会汇总得到的 value,最后的结果的条数是根据 key 决定的,有多少 key 就汇总成多少条数据,然后把 RDD 的所有相同 key 的 value 放到一个 Iterable 里面,类似于类似 SQL 里的全连接。代码示例如下:

```scala
import org.apache.spark.{SparkConf,SparkContext}
object cogroupAndjoin{
  def main(args: Array[String]){
    //创建 SparkConf 对象
    val conf = new SparkConf().setAppName("cogroupAndjoin").setMaster("local")
    //创建 SparkContext 对象
    val sc = new SparkContext(conf)
    //创建两个可被并行操作的分布式数据集
    val idName = sc.parallelize(Array((1,"张三"), (2,"李四"), (3,"王五")))
    val idAge = sc.parallelize(Array((1, 30), (2,29), (4,21)))
    println("\ncogroup\n")
    //对两个并行数据集进行 cogroup 操作
    idName.cogroup(idAge).collect().foreach(println)
    println("\njoin\n")
    //对两个并行数据集进行 join 操作
```

```
      idName.join(idAge).collect().foreach(println)
  }
}
```

2. 其他常用算子

新建 Scala 类 NoStateStreamDemo,演示无状态转换操作其他相应算子,输入流采用 Socket 套接字流,示例代码和说明如下:

```
import org.apache.spark._
import org.apache.spark.streaming._
object NoStateStreamDemo{
  def main(args: Array[String]){
    //创建 SparkConf 对象
    val sparkConf = new SparkConf().setAppName("NoStateStreamDemo").setMaster("local[2]")
    //创建 StreamingContext 对象,时间间隔为 5s
    val ssc = new StreamingContext(sparkConf,Seconds(5))
    //获取 Socket 数据,端口号为 9999
    var lines = ssc.socketTextStream("hhaonote", 9999)
    //将 lines 根据空格进行分割,分割成若干个单词
    val words = lines.flatMap(_.split(" "))
    //将每一个单词都用"-"符号进行拼接
    val ReduceWords = words.Reduce(_ + "-" + _)
    //将每一个单词后面都加上"_one"
    val wordsOne = words.Map(_ + "_one")
    //将每一个单词后面都加上"_two"
    val wordsTwo = words.Map(_ + "_two")
    //生成一个新的 DStream,只包含一个元素,对应语句单词统计数值
    val wordsCount = words.count()
    //统计 words 中不同单词的个数
    val countByValueWords = words.countByValue()
    //对 words 进行判断,去除 hello 这个单词
    val filterWords = words.filter(_ != "hello")
    //将 wordsOne 和 wordsTwo 两个 RDD 进行合并
    val unionWords = wordsOne.union(wordsTwo)
    //打印输出
    wordsOne.print()
    wordsTwo.print()
    wordsCount.print()
    unionWords.print()
    ReduceWords.print()
    countByValueWords.print()
    filterWords.print()
    ssc.start()
```

```
        ssc.awaitTermination()
    }
}
```

在 Windows 操作系统中安装 Netcat，在命令行输入 nc -l -p 9999 命令（Linux 操作系统命令为 nc -lk 9999），输入下面两行字符串：

```
hello hadoop
hello spark
```

每一个单词后面加"_one"运行结果如图 10-5 所示。每一个单词后面加"_two"运行结果如图 10-6 所示。对应语句单词统计数如图 10-7 所示。两个 RDD 合并如图 10-8 所示。每一个单词用"-"符号进行拼接如图 10-9 所示。统计不同单词的个数如图 10-10 所示。去除 hello 运行结果如图 10-11 所示。

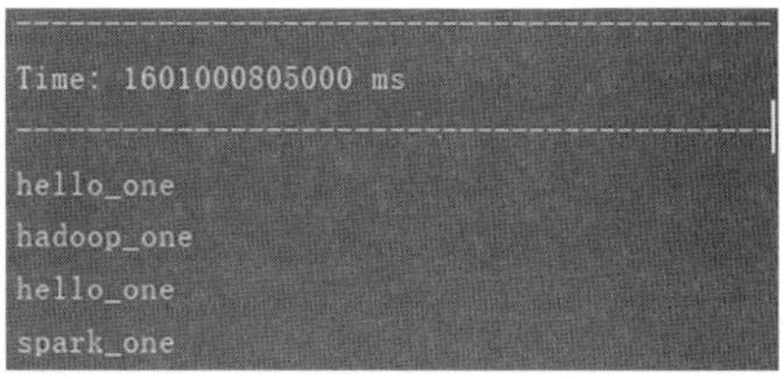

图 10-5 每一个单词后面加"_one"运行结果

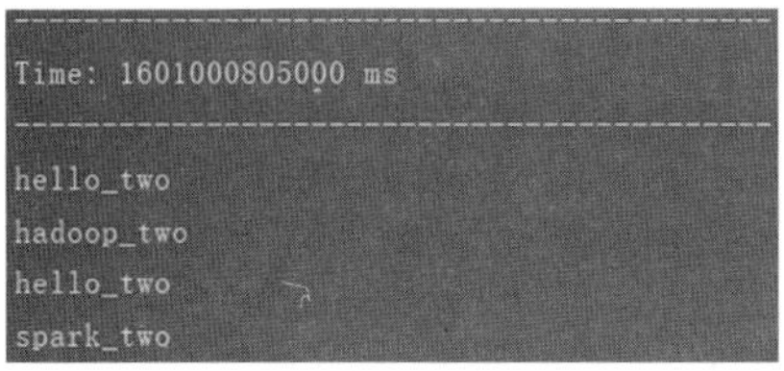

图 10-6 每一个单词后面加"_two"运行结果

图 10-7 对应语句单词统计数

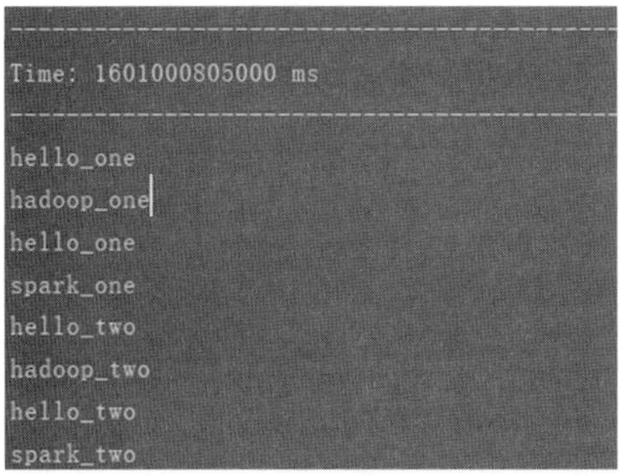

图 10-8 两个 RDD 合并

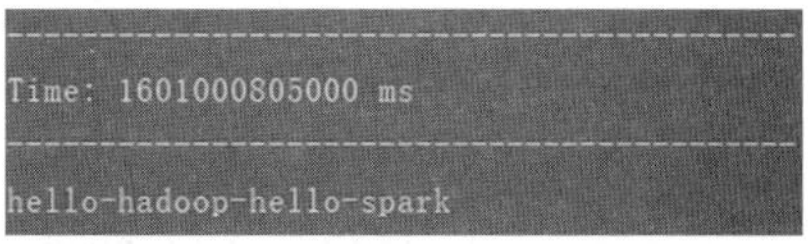

图 10-9 每一个单词用"-"符号进行拼接

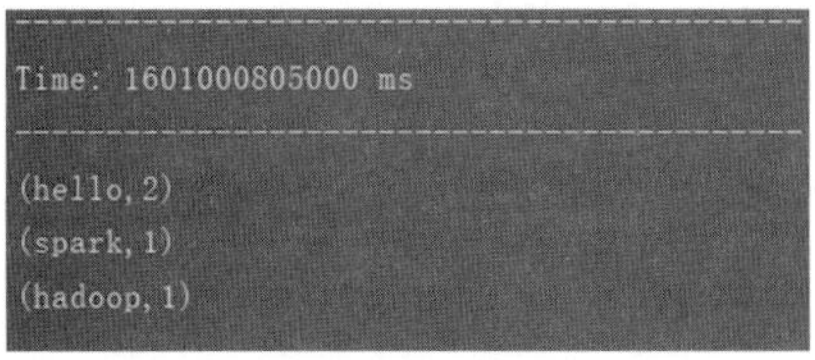

图 10-10 统计不同单词的个数

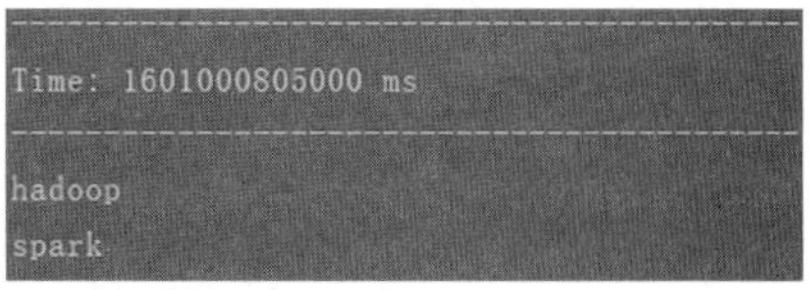

图 10-11 去除 hello 运行结果

10.3.3 有状态转换操作

1. 滑动窗口

滑动窗口转换操作的计算过程如图 10-12 所示。事先设定一个滑动窗口的长度（也就是窗口的持续时间），并且设定滑动窗口的时间间隔（每隔多长时间执行一次计算），然后让窗口按照指定时间间隔在源 DStream 上滑动，每次窗口停放的位置上都会有一部分 DStream 被框入窗口内，形成一个小段的 DStream，可以启动对这个小段 DStream 的计算。

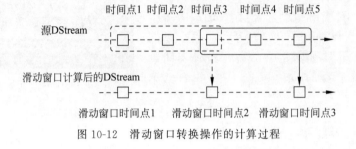

图 10-12　滑动窗口转换操作的计算过程

为防止数据重复使用或丢失，要保证窗口的长度与滑动时间相同，同时要保证窗口的长度与滑动时间是最小批次的整数倍。示例代码及说明如下：

```
import org.apache.spark.streaming.dstream.{DStream,ReceiverInputDStream}
import org.apache.spark.streaming.{Seconds,StreamingContext}
import org.apache.spark.{SparkConf,SparkContext}
object SparkStreamingScoketWindow{
  def main(args: Array[String]): Unit = {
    //创建 SparkConf 对象
    val sparkConf: SparkConf = new SparkConf().setAppName("SparkStreamingScoketWindow").setMaster("local[2]")
    //创建 SparkContext 对象
    val sparkContext = new SparkContext(sparkConf)
    //设置日志级别
    sparkContext.setLogLevel("warn")
    //获取 StreamingContext 对象,5s 一个批次
    val ssc = new StreamingContext(sparkContex,Seconds(5))
    //接收 Socket 的数据
     val textStream: ReceiverInputDStream[String] = ssc.socketTextStream("hhaonote",9999)
    //获取每一行的单词
    val words: DStream[String] = textStream.flatMap(_.split(" "))
    //将每一个单词置为 1
    val wordAndOne: DStream[(String,Int)] = words.Map((_,1))
     //每隔 10s 统计最近 10s 的搜索词出现的次数
    val result: DStream[(String,Int)] = wordAndOne.ReduceByKeyAndWindow((x:Int,y:Int) => x + y,Seconds(10),Seconds(10))
    //打印
    result.print()
```

```
        //开启流式处理
        ssc.start()
        ssc.awaitTermination()
    }
  }
}
```

在 Windows 操作系统中安装 Netcat，在命令行输入 nc -l -p 9999 命令（Linux 操作系统上命令为 nc -lk 9999），不断地输入字符串，可以看到程序每隔 10s 统计最近 20s 的搜索词出现的次数。

2. 状态管理函数 updateStateByKey()

Spark Streaming 的 updateStateByKey() 可以通过设定检查点记录历史批次状态，通过定义状态更新函数，在对 DStream 中的数据进行按 key 做 Reduce 操作后，可以对各个批次的数据进行累加或其他操作。示例代码及说明如下：

```
import org.apache.spark.{SparkConf,SparkContext}
import org.apache.spark.streaming.{Seconds,StreamingContext}
object updateStateByKeySocketDemo{
    def main(args: Array[String]): Unit = {
        //定义状态更新函数
        val updateFunc = (currValues: Seq[Int],prevValueState: Option[Int]) => {
        //返回结果对象（迭代值）初始值为 0，累计 currentCount 集合中的每个值
        val currentCount = currValues.foldLeft(0)(_ + _)
        //已累加的值
        val previousCount = prevValueState.getOrElse(0)
        //返回累加后的结果，是一个 Option[Int]类型
        Some(currentCount + previousCount)
        }
        //创建 SparkConf 对象
        val sparkConf = new SparkConf().setAppName("updateStateByKeySocketDemo")
        .setMaster("local[2]")
        //创建 SparkContext 对象
        val sparkContext = new SparkContext(sparkConf)
        //获取 StreamingContext 对象，5s 一个批次
        val ssc = new StreamingContext(sparkContext,Seconds(5))
        //设置日志级别
        sparkContext.setLogLevel("warn")
        //开启检查点
        ssc.checkpoint("E:/checkpoint")
        //接收数据源，产生 DStream 对象
        val lines = ssc.socketTextStream("hhaonote",9999)
        //以空格分割单词
        val words = lines.flatMap(_.split(" "))
        //将分割后的每个单词出现次数记录为 1
        val wordDstream = words.Map(x => (x, 1))
        //通过键的先前值和新值上应用给定函数 updateFunc 更新每一个键的状态
        val stateDstream = wordDstream.updateStateByKey[Int](updateFunc)
```

```
            //输出打印
            stateDstream.print()
            //开启实时计算
            ssc.start()
            //等待应用停止
            ssc.awaitTermination()
        }
    }
```

在Windows操作系统的命令行输入 nc -l -p 9999 命令(Linux操作系统上命令为 nc -lk 9999),不断输入字符串,可以看到单词个数是以往批次和当前批次的累加值, updateStateByKey示例运行结果如图10-13所示。

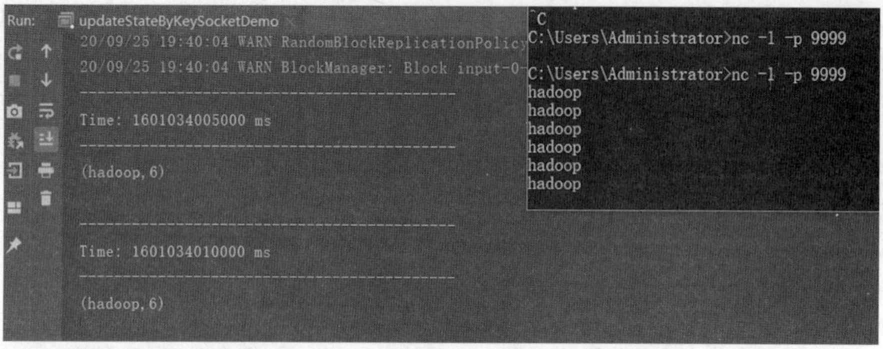

图10-13　updateStateByKey()示例运行结果

10.3.4　输出操作

将 saveAsTextFiles() 代码放入 updateStateByKeySocketDemo.scala 代码中,修改后的代码如下:

```
import org.apache.spark.{SparkConf,SparkContext}
import org.apache.spark.streaming.{Seconds,StreamingContext}
object updateStateByKeySocketDemo{
  def main(args: Array[String]): Unit = {
    //定义状态更新函数
    val updateFunc = (currValues: Seq[Int], prevValueState: Option[Int]) => {
    //返回结果对象(迭代值)初始值为0,累计currentCount集合中的每个值
    val currentCount = currValues.foldLeft(0)(_ + _)
    //已累加的值
    val previousCount = prevValueState.getOrElse(0)
    //返回累加后的结果,是一个Option[Int]类型
    Some(currentCount + previousCount)
    }
    //创建SparkConf对象
    val sparkConf = newSparkConf().setAppName("updateStateByKeySocketDemo")
    .setMaster("local[2]")
    //创建SparkContext对象
```

```
    val sparkContext = new SparkContext(sparkConf)
    //获取 StreamingContext 对象,5s 一个批次
    val ssc = new StreamingContext(sparkContext,Seconds(5))
    //设置日志级别
    sparkContext.setLogLevel("warn")
    //开启检查点
    ssc.checkpoint("E:/checkpoint")
    //接收数据源,产生 DStream 对象
    val lines = ssc.socketTextStream("hhaonote",9999)
    //以空格分割单词
    val words = lines.flatMap(_.split(" "))
    //将分割后的某个单词每出现次数记录为 1
    val wordDstream = words.Map(x => (x, 1))
    //通过键的先前值和在新值上应用给定函数 updateFunc()更新每一个键的状态
    val stateDstream = wordDstream.updateStateByKey[Int](updateFunc)
    //输出打印
    stateDstream.print()
    //新加的代码,把 DStream 保存到文本文件中
    stateDstream.saveAsTextFiles("E:/log/output.txt")
    //开启实时计算
    ssc.start()
    //等待应用停止
    ssc.awaitTermination()
  }
}
```

运行程序,输出操作示例运行结果,如图 10-14 所示。可以发现,在 E 盘 log 目录下,生成了很多文本文件。由于在代码中有一句"val ssc = new StreamingContext(sparkContext, Seconds(5));",也就是说,每隔 5s 统计一次词频,所以,每隔 5s 就会生成一次词频统计结果,并输出到 E 盘 log 目录的 output.txt-时间标记下(如 1601035215000)。

名称	修改日期	类型
output.txt-1601035215000	2020-9-25 20:00	文件夹
output.txt-1601035220000	2020-9-25 20:00	文件夹
output.txt-1601035225000	2020-9-25 20:00	文件夹
output.txt-1601035230000	2020-9-25 20:00	文件夹
output.txt-1601035235000	2020-9-25 20:00	文件夹
output.txt-1601035240000	2020-9-25 20:00	文件夹
output.txt-1601035245000	2020-9-25 20:00	文件夹
output.txt-1601035250000	2020-9-25 20:00	文件夹
output.txt-1601035255000	2020-9-25 20:00	文件夹
output.txt-1601035260000	2020-9-25 20:01	文件夹
output.txt-1601035265000	2020-9-25 20:01	文件夹

图 10-14 输出操作示例运行结果

10.4　本章小结

本章主要介绍了 Spark Streaming 的一些基本概念和原理,介绍了 DStream 编程模型,最后通过三个实例演示了用 Scala 编程实现 DStream 的有状态转换操作、无状态转换操作、输出操作,以加深读者对本章的理解。

第 11 章

Spark Streaming 与 Flume、Kafka 的整合

11.1 Flume 简介及安装

11.1.1 Flume 简介

Flume 是 Cloudera 提供的一个高可用、高可靠、分布式的海量日志采集、聚合和传输的系统。Flume 支持在日志系统中定制各类数据发送方,用于收集数据;同时,Flume 提供对数据进行简单处理,并将数据写到各种数据接收方(可定制)的能力。

Flume 主要由以下三个重要的组件构成。

(1) Source:完成对日志数据的收集,并将收集数据汇聚到 Channel 中。

(2) Channel:主要提供一个队列的功能,对 Source 提供的数据进行简单的缓存。

(3) Sink:取出 Channel 中的数据,存储到文件系统、数据库,或者提交到远程服务器。

Flume 组件如图 11-1 所示。

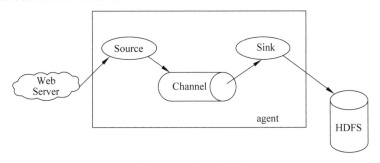

图 11-1 Flume 组件

Flume 逻辑上分三层架构：agent、collector、storage。agent 用于采集数据，agent 是 Flume 中产生数据流的地方，同时，agent 会将产生的数据流传输到 collector。collector 的作用是将多个 agent 的数据汇总后，加载到 storage 中。storage 是存储系统，可以是一个普通文件，也可以是 HDFS、Hive、HBase 等。Flume 逻辑架构如图 11-2 所示。

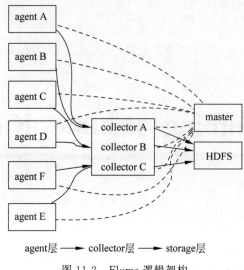

图 11-2 Flume 逻辑架构

11.1.2 Flume 的安装

Flume 的安装步骤如下。

（1）下载 apache-flume-1.8.0-bin.tar.gz，下载网址是 https://mirrors.bfsu.edu.cn/apache/flume/1.8.0/apache-flume-1.8.0-bin.tar.gz。

（2）解压安装包，执行 CentOS 命令 tar -zxvf apache-flume-1.8.0-bin.tar.gz -C /hadoop/，进行解压缩，解压路径是 /hadoop。

（3）编辑 /etc/profile 文件，加入以下内容：

```
export FLUME_HOME = /hadoop/apache-flume-1.8.0-bin
export FLUME_CONF_DIR = $FLUME_HOME/conf
export PATH = $PATH:$FLUME_HOME/bin
```

保存文件并退出。输入 CentOS 命令 source /etc/profile，使配置生效。

（4）编辑 flume-env.sh 文件，将 conf 路径下的 flume-env.sh.template 文件更名为 flume-env.sh，在文件的最开始位置增加 JAVA_HOME 设置。

```
export  JAVA_HOME = /usr/java/jdk1.8.0_161
```

（5）编辑 flume-conf.properties 文件，将 conf 路径下的 flume-conf.template 文件更名为 flume-conf，在文件末尾加入以下内容：

```
#agent1 表示代理名称
agent1.sources = source1
```

```
agent1.sinks = sink1
agent1.channels = channel1
#配置 source1
agent1.sources.source1.type = spooldir
agent1.sources.source1.spoolDir = /hadoop/apache-flume-1.8.0-bin/logs
agent1.sources.source1.channels = channel1
agent1.sources.source1.fileHeader = false
agent1.sources.source1.interceptors = i1
agent1.sources.source1.interceptors.i1.type = timestamp
#配置 channel1
agent1.channels.channel1.type = file
agent1.channels.channel1.checkpointDir = /hadoop/apache-flume-1.8.0-bin/logs_tmp_cp
agent1.channels.channel1.dataDirs = /hadoop/apache-flume-1.8.0-bin/logs_tmp
#配置 sink1
agent1.sinks.sink1.type = hdfs
agent1.sinks.sink1.hdfs.path = hdfs://bigdata01:9000/flumelogs
agent1.sinks.sink1.hdfs.fileType = DataStream
agent1.sinks.sink1.hdfs.writeFormat = TEXT
agent1.sinks.sink1.hdfs.rollInterval = 1
agent1.sinks.sink1.channel = channel1
agent1.sinks.sink1.hdfs.filePrefix = %Y-%m-%d
```

配置文件中 agent1.sources.source1.spoolDir 配置表示监听的文件夹,agent1.sinks.sink1.hdfs.path 配置表示监听到的文件自动上传到 HDFS 的路径,需要手动创建 HDFS 下的目录。相应 HDFS 命令如下：hdfs dfs -mkdir /flumelogs。

(6) 启动服务,执行地 Flume 命令：

```
flume-ng agent -n agent1 -c conf -f
    /hadoop/apache-flume-1.8.0-bin/conf/flume-conf.properties -Dflume.root.logger
    =DEBUG,console
```

(7) 测试,在/hadoop/apache-flume-1.8.0-bin/logs 创建一个 test.txt 文件,输入文本 hello flume,可以看到 Flume 的 agent 自动上传了刚刚创建的文件。Flume 安装测试如图 11-3 所示。

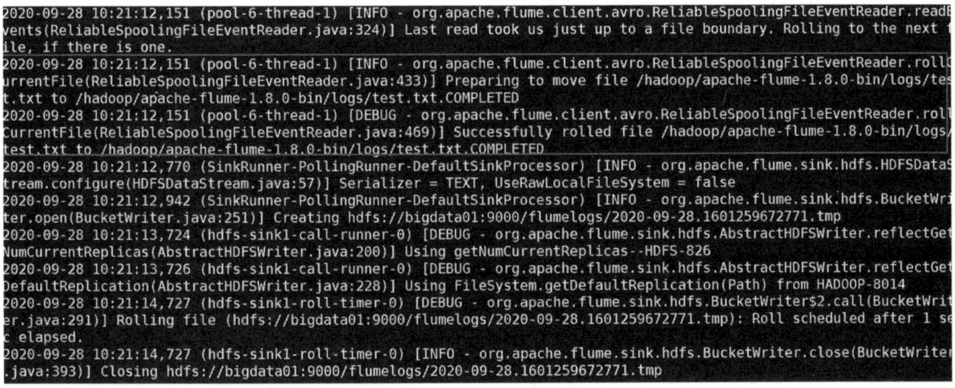

图 11-3　Flume 安装测试

11.2 Kafka 简介及安装

11.2.1 Kafka 简介

Kafka 是一种高吞吐量的分布式发布订阅消息系统,它可以处理生产者和消费者的所有动作流数据。生产者（Producer）向 Kafka 集群发送消息,在发送消息之前,会对消息进行分类,即主题(Topic),通过对消息指定主题可以将消息分类,消费者可以只关注自己需要的 Topic 中的消息。消费者(Consumer)通过与 Kafka 集群建立长连接的方式,不断地从集群中接收消息,然后可以对这些消息进行处理。Kafka 消息处理框架图如图 11-4 所示。

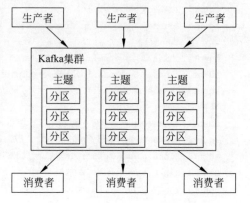

图 11-4　Kafka 消息处理框架图

Kafka 相关概念如下。

(1) Broker。Kafka 集群包含一个或多个服务器,这种服务器被称为 Broker。

(2) Topic：消息主题。每条发布到 Kafka 集群的消息都有一个类别,这个类别被称为 Topic。物理上不同 Topic 的消息分开存储；逻辑上一个 Topic 的消息虽然保存于一个或多个 Broker 上,但用户只需指定消息的 Topic 即可生产或消费数据而不必关心数据存于何处。

(3) Partition：分区。Partition 是物理上的概念,每个 Topic 包含一个或多个 Partition。

(4) Producer：生产者。负责发布消息到 Kafka Broker。

(5) Consumer：消费者。读取 Kafka Broker 消息的客户端。

(6) Consumer Group。每个 Consumer 都属于一个特定的组(可为每个消费者指定组名,若不指定组名则属于默认的组)。

11.2.2 Kafka 的安装

Kafka 的安装步骤如下。

(1) 下载/Kafka 安装包,下载网址是 https://mirror.bit.edu.cn/apache/kafka/2.4.1/kafka_2.11-2.4.1.tgz。

(2) 解压缩安装包,执行 CentOS 命令 tar -zxvf kafka_2.11－2.4.1.tgz -C /hadoop/进行解压缩,解压路径是/hadoop。

(3) 编辑/etc/profile 文件,加入以下内容：

```
export ZOOKEEPER_HOME = /hadoop/zookeeper－3.4.13
export KAFKA_HOME = /hadoop/kafka_2.11－2.4.1
export PATH = $PATH:$KAFKA_HOME/bin
```

保存文件并退出。输入 CentOS 命令 source /etc/profile，使配置生效。

(4) 编辑 server.properties 文件，加入以下内容：

```
broker.id = 0
listeners = PLAINTEXT://172.16.106.69:9092
advertised.listeners = PLAINTEXT://172.16.106.69:9092
zookeeper.connect = 172.16.106.69:2181,172.16.106.70:2181,172.16.106.71:2181
```

(5) 将安装文件复制到另两个节点，CentOS 命令如下：

```
scp -r kafka_2.11-2.4.1 root@172.16.106.70:/hadoop/
scp -r kafka_2.11-2.4.1 root@172.16.106.71:/hadoop/
```

(6) 修改另两个节点上的相关配置，内容如下。

① 修改 server.properties 文件，内容如下。

将 broker.id 分别修改成 1 或 2。

将 listeners 在 IP 处分别修改成子节点对应的 IP。

将 advertised.listeners 在 IP 处分别修改成子节点对应 IP。

② 修改另外两个节点的/etc/profile 文件，配置 Kafka 环境变量。

(7) 启动服务，在三个节点都启动，在 Kafka 上执行以下命令：

```
./kafka-server-start.sh /hadoop/kafka_2.11-2.4.1/config/server.properties &
```

(8) 测试，在主节点上创建主题 TestTopic，命令如下：

```
kafka-topics.sh
-- zookeeper 172.16.106.69:2181,172.16.106.70:2181,172.16.106.71:2181
-- topic TestTopic -- replication-factor 1 -- partitions 1 -- create
```

在主节点上启动一个生产者，命令如下：

```
kafka-console-producer.sh
-- broker-list 172.16.106.69:9092,172.16.106.70:9092,172.16.106.71:9092
-- topic TestTopic
```

在其他两个节点上分别创建消费，命令如下：

```
kafka-console-consumer.sh
-- bootstrap-server 172.16.106.70:9092 -- topic TestTopic -- from-beginning
kafka-console-consumer.sh
-- bootstrap-server 172.16.106.71:9092 -- topic TestTopic -- from-beginning
```

在主节点生产者命令行输入一段文字，生产者测试结果如图 11-5 所示。

图 11-5　生产者测试结果

在两个消费者节点上，可以看到生产者的输入文字，表示消费者收到了该数据。消费者测试结果如图 11-6 所示。

```
[2020-09-28 11:42:27,771] INFO [GroupCoordinator 2]: Stabilized group console-consumer-54312 generation 1 (__consumer_o
ffsets-26) (kafka.coordinator.group.GroupCoordinator)
[2020-09-28 11:42:27,779] INFO [GroupCoordinator 2]: Assignment received from leader for group console-consumer-54312 f
or generation 1 (kafka.coordinator.group.GroupCoordinator)
hello wor
hello
hello word
hello world
```

图 11-6　消费者测试结果

查看 HDFS 下 flumelogs 下的文件，可以看到文件已经自动上传到 HDFS 中，文件命名方式是按 conf 文件中的 filePrefix 的配置格式。HDFS 测试结果如图 11-7 所示。

```
[root@bigdata01 logs]# hdfs dfs -ls -R /flumelogs
-rw-r--r--   1 root supergroup         12 2020-09-28 10:21 /flumelogs/2020-09-28.1601259672771
[root@bigdata01 logs]# hdfs dfs -cat /flumelogs/2020-09-28.1601259672771
hello flume
[root@bigdata01 logs]#
```

图 11-7　HDFS 测试结果

11.3　Flume 与 Kafka 的区别和侧重点

(1) Kafka 是一个非常通用的系统，可以有许多生产者和消费者共享多个主题。相比之下，Flume 是一个向 HDFS、HBase 等发送数据的专用工具。它对 HDFS 有特殊的优化，并且集成了 Hadoop 的安全特性。如果数据被多个系统消费，则使用 Kafka；如果数据有多个生产者场景，或者有写入 Hbase、HDFS 操作，则使用 Flume。

(2) Flume 可以使用拦截器实时处理数据。而 Kafka 需要外部的流处理系统才能做到。

(3) Kafka 和 Flume 都是可靠的系统，通过适当的配置能保证零数据丢失。然而，Flume 不支持副本事件。如果 Flume 代理的一个节点崩溃了，即使使用了可靠的文件管道方式，也将丢失这些事件直到恢复这些磁盘。如果需要一个高可靠性的管道，那么使用 Kafka 是一个更好的选择。

目前，Streaming＋Flume＋Kafka 是大数据准实时数据采集的最为可靠并且也是最常用的方案，三者可以协作工作，主要有以下两种方式。

(1) Kafka＋Flume。如果需要把流式数据从 Kafka 转移到 Hadoop，可以使用 Flume 代理(Agent)，将 Kafka 当作一个来源(Source)，这样可以从 Kafka 读取数据到 Hadoop。

(2) Flume＋Kafka。做日志缓存，Flume 的数据采集部分做得很好，可以定制很多数据源，减少开发量，然后利用 Kafka 分发到多个系统。

11.4　Spark Streaming 与 Flume、Kafka 的整合与开发

以一个示例说明 Spark Streaming 与 Flume、Kafka 的整合与开发过程，示例的功能是商品实时交易数据统计分析，通过 Flume 实时收集交易订单，将数据分发到 Kafka，

Kafka 将数据传输到 Spark Streaming,Spark Streaming 统计商品的销售量。主要有以下几个步骤。

(1) 通过日志模拟程序产生实时交易数据。
(2) Flume 收集模拟产生的实时交易数据。
(3) Flume 将数据发送给 Kafka 消息队列。
(4) Spark Streaming 接收 Kafka 消息队列的消息,每 5s 进行一次数据统计。

具体实现如下。

(1) 新建 MAVEN 项目,项目名称为 RealtimeAnalysis,新建过程见第 9 章。在 pom.xml 文件中加入依赖包,代码如下:

```xml
<?xml version="1.0" encoding="UTF-8"?>
<project xmlns="http://maven.apache.org/POM/4.0.0"
    xmlns:xsi="http://www.w3.org/2001/XMLSchema-instance"
    xsi:schemaLocation="http://maven.apache.org/POM/4.0.0
    http://maven.apache.org/xsd/maven-4.0.0.xsd">
    <modelVersion>4.0.0</modelVersion>
    <groupId>com.sparklearn</groupId>
    <artifactId>RealtimeAnalysis</artifactId>
    <version>1.0-SNAPSHOT</version>
    <properties>
        <spark.version>2.1.0</spark.version>
        <scala.version>2.11.8</scala.version>
    </properties>
    <dependencies>
        <dependency>
            <groupId>org.apache.spark</groupId>
            <artifactId>spark-core_2.11</artifactId>
            <version>${spark.version}</version>
        </dependency>
        <dependency>
            <groupId>org.apache.spark</groupId>
            <artifactId>spark-streaming_2.11</artifactId>
            <version>${spark.version}</version>
        </dependency>
        <dependency>
            <groupId>org.apache.spark</groupId>
            <artifactId>spark-mllib_2.11</artifactId>
            <version>${spark.version}</version>
        </dependency>
        <dependency>
            <groupId>org.apache.spark</groupId>
            <artifactId>spark-streaming-kafka-0-8_2.11</artifactId>
            <version>${spark.version}</version>
        </dependency>
        <!-- Kafka -->
        <dependency>
            <groupId>org.apache.flume.flume-ng-clients</groupId>
```

```xml
            <artifactId>flume-ng-log4jappender</artifactId>
            <version>1.6.0</version>
        </dependency>
        <dependency>
            <groupId>log4j</groupId>
            <artifactId>log4j</artifactId>
            <version>1.2.17</version>
        </dependency>
        <dependency>
            <groupId>org.apache.flume</groupId>
            <artifactId>flume-ng-core</artifactId>
            <version>1.6.0</version>
        </dependency>
        <dependency>
            <groupId>org.apache.kafka</groupId>
            <artifactId>kafka_2.10</artifactId>
            <version>0.8.2.1</version>
        </dependency>
        <!-- JSON 数据转换工具 -->
        <dependency>
            <groupId>com.alibaba</groupId>
            <artifactId>fastjson</artifactId>
            <version>1.2.41</version>
        </dependency>
    </dependencies>
</project>
```

（2）新建 log4j.properties 文件。在工程的 resource 目录下新建 log4j.properties 文件，内容如下（其中注意 log4j.appender.flume.Hostname 的配置，要配置成安装 Flume 的服务器）：

```
log4j.rootLogger = INFO,stdout,flume
log4j.appender.stdout = org.apache.log4j.ConsoleAppender
log4j.appender.stdout.target = System.out
log4j.appender.stdout.layout = org.apache.log4j.PatternLayout
log4j.appender.stdout.layout.ConversionPattern = %d{yyyy-MM-dd HH:mm:ss,SSS} [%t] [%c] [%p] - %m%n
log4j.appender.flume = org.apache.flume.clients.log4jappender.Log4jAppender
log4j.appender.flume.Hostname = 172.16.106.69
log4j.appender.flume.Port = 41414
log4j.appender.flume.UnsafeMode = true
```

（3）新建 Java 类 LoggerGenerator。在工程的 test 目录下新建 Java 类 LoggerGenerator，此类用于不断模拟产生订单交易数据，在此类中每 6s 调用一次 PaymentInfo 交易的实体类的 random()方法是模拟产生订单交易数据方法，数据以 JSON 格式返回。其中，PaymentInfo 是交易的实体类，有三个成员变量，分别是订单编号、商品价格、商品编号；LoggerGenerator 为模拟日志生成类。相应代码如下：

```java
import com.alibaba.fastjson.JSONObject;
```

```java
import java.util.Random;
import java.util.UUID;
public class PaymentInfo{
    private static final long serialVersionUID = 1L;
    private String orderId;//订单编号
    private String productId;//商品编号
    private long productPrice;//商品价格
    public PaymentInfo(){
    }
    public static long getSerialVersionUID(){
        return serialVersionUID;
    }
    public String getOrderId(){
        return orderId;
    }
    public void setOrderId(String orderId){
        this.orderId = orderId;
    }
    public String getProductId(){
        return productId;
    }
    public void setProductId(String productId){
        this.productId = productId;
    }
    public long getProductPrice(){
        return productPrice;
    }
    public void setProductPrice(long productPrice){
        this.productPrice = productPrice;
    }
    @Override
    public String toString(){
        return "PaymentInfo{" +
                "orderId='" + orderId + '\'' +
                ", productId='" + productId + '\'' +
                ", productPrice=" + productPrice +
                '}';
    }
    //随机模拟订单数据
    public String random(){
    //创建随机对象
        Random r = new Random();
        //生成订单编号
        this.orderId = UUID.randomUUID().toString().replaceAll( "-", "" );
        //生成商品价格
        this.productPrice = r.nextInt( 1000 );
        //生成商品编号
        this.productId = r.nextInt( 10 ) + "";
        //转换为JSON格式
```

```java
            JSONObject obj = new JSONObject();
            String jsonString = obj.toJSONString( this );
            return jsonString;
        }
    }


import org.apache.log4j.Logger;
public class LoggerGenerator{
    private static Logger logger = Logger.getLogger( LoggerGenerator.class );
    public static void main(String[] args){
        while (true){

            PaymentInfo p = new PaymentInfo();
            //随机模拟订单数据
            String message = p.random();
            logger.info( message );
            try {
               //延时 6s
               Thread.sleep( 6000 );
            } catch (InterruptedException e) {
                //TODO Auto-generated catch block
                e.printStackTrace();
            }
        }
    }
}
```

（4）新建 log4j_flume.properties 文件。在安装 Flume 服务器的 conf 目录下新建文件 log4j_flume.properties，其中注意 sinks.kafka_sink.brokerList 配置的是连接 Kafka 集群的地址和端口号。内容如下：

```
log4j_agent.sources = avro_source
log4j_agent.channels = memory_channel
log4j_agent.sinks = kafka_sink
#配置 source
log4j_agent.sources.avro_source.type = avro
log4j_agent.sources.avro_source.bind = 0.0.0.0
log4j_agent.sources.avro_source.port = 41414
#配置 channel
log4j_agent.channels.memory_channel.type = memory
log4j_agent.channels.memory_channel.capacity = 10000
log4j_agent.channels.memory_channel.transactionCapacity = 10000
#配置 sink,并连接到 Kafka
log4j_agent.sinks.kafka_sink.type = org.apache.flume.sink.kafka.KafkaSink
log4j_agent.sinks.kafka_sink.topic = logtoflume
log4j_agent.sinks.kafka_sink.brokerList = 172.16.106.69:9092,172.16.106.70:9092,172.16.106.71:9092
log4j_agent.sinks.kafka_sink.requiredAcks = 1
log4j_agent.sinks.kafka_sink.batchSize = 20
```

```
log4j_agent.sources.avro_source.channels = memory_channel
log4j_agent.sinks.kafka_sink.channel = memory_channel
```

(5) 启动 Flume,命令如下:

```
flume-ng agent -n agent1 -c conf -f /hadoop/apache-flume-1.8.0-bin/conf/log4j_flume.properties -Dflume.root.logger=DEBUG,console
```

(6) 启动 Kafka 并新建 Topic,名称为 logtoflume,命令如下:

```
./kafka-server-start.sh /hadoop/kafka_2.11-2.4.1/config/server.properties &
kafka-topics.sh
--zookeeper 172.16.106.69:2181,172.16.106.70:2181,172.16.106.71:2181
--topic logtoflume --replication-factor 1 --partitions 1 --create
```

(7) 新建 Scala 类 KafkaConsumerMsg,接收 Kafka 下的名称为 logtoflumeTopic 的队列数据,并做统计,代码及说明如下:

```scala
import kafka.serializer.StringDecoder
import com.alibaba.fastjson.{JSON, JSONObject}
import org.apache.spark.streaming.dstream.{DStream, InputDStream}
import org.apache.spark.streaming.kafka.KafkaUtils
import org.apache.spark.streaming.{Seconds, StreamingContext}
import org.apache.spark.{SparkConf, SparkContext}
object KafkaConsumerMsg{
  def main(args: Array[String]): Unit = {
    //创建 SparkConf 对象
    val sparkConf = new SparkConf().setMaster("local[2]").setAppName("KafkaConsumerMsg")
    //创建 SparkContex 对象
    val sparkContext = new SparkContext(sparkConf)
    //设置日志级别
    sparkContext.setLogLevel("warn")
    //获取 StreamingContext 对象,5s 一个批次
    val ssc = new StreamingContext(sparkContext,Seconds(5))
    //设置 Kafka 参数
    val kafkaParams = Map("bootstrap.servers" -> "bigdata01:9092,bigdata02:9092,bigdata03:9092", "group.id" -> "spark-receiver")
    //指定 Topic 相关信息
    val topics = Set("logtoflume")
    //通过 KafkaUtils.createDirectStream 利用低级 API 接收 Kafka 数据
    val kafkaDstream: InputDStream[(String, String)] =
      KafkaUtils.createDirectStream
      [String, String, StringDecoder,StringDecoder](ssc, kafkaParams,topics)
    //获取 Kafka 中 Topic 数据,并解析 JSON 格式数据
    val events: DStream[JSONObject] = kafkaDstream.flatMap(line => Some(JSON.parseObject(line._2)))
    //按照 productId 进行分组统计个数和总价格
    val orders: DStream[(String, Int, Long)] = events.Map(x => (x.getString("productId"),x.getLong("productPrice"))).qroupByKey() Map(x -> (x._1, x._2.size, x._2.ReduceLeft(_ + _)))
    //打印输出
    orders.foreachRDD(x =>
```

```
                x.foreachPartition(partition =>
                    partition.foreach(x => {
                        println("productId = "
                            + x._1 + " count = " + x._2 + " productPricrice = " + x._3)
                    })
                )
            )
        ssc.start()
        ssc.awaitTermination()
        }
    }
```

(8) 启动 LoggerGenerator, 模拟产生订单交易数据。LoggerGenerator 运行界面如图 11-8 所示。

图 11-8 LoggerGenerator 运行界面

(9) 启动 KafkaConsumerMsg 接收 Kafka 下的 Topic 队列的数据, 并做统计。KafkaConsumerMsg 运行界面如图 11-9 所示。

图 11-9 KafkaConsumerMsg 运行界面

11.5 本章小结

本章主要介绍了 Spark Streaming 与 Flume、Kafka 的整合, 介绍了 Flume 和 Kafka 的安装过程, 最后通过一个例子演示了用 Scala 编程实现 Spark Streaming 与 Flume 和 Kafka 的整合。

第 12 章

Spark MLlib机器学习

12.1 机器学习的概念

12.1.1 机器学习的定义

机器学习是一种通过利用数据,训练出模型,然后使用模型预测的一种方法。机器学习的构建过程是,利用数据通过算法构建出模型并对模型进行评估,评估的性能如果达到要求就拿这个模型来测试其他数据,如果达不到要求就调整算法来重新建立模型,再次进行评估,如此循环往复,最终获得满意的方法来处理其他数据。机器学习模型构建过程如图 12-1 所示。

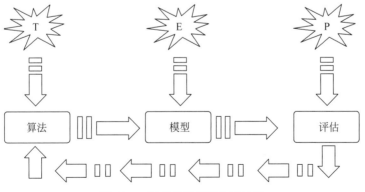

图 12-1 机器学习模型构建过程

12.1.2 机器学习的分类

1. 监督学习

监督学习是从给定的训练数据集中学习一个函数(模型),当新的数据到来时,可以根据这个函数(模型)预测结果。监督学习的训练集要求包括输入和输出,也可以说是特征和目标。训练集中的目标是由人标注的。在监督学习下,输入数据被称为"训练数据",每组训练数据有一个明确的标识或结果,如在防垃圾邮件系统中标识"垃圾邮件""非垃圾邮件",在手写数字识别中标识1、2、3等。在建立预测模型时,监督学习建立一个学习过程,将预测结果与"训练数据"的实际结果进行比较,不断调整预测模型,直到模型的预测结果达到一个预期的准确率。常见的监督学习算法包括回归分析和分类。

2. 无监督学习

与监督学习相比,无监督学习的训练集没有人为标注的结果。在非监督学习中,数据并不被特别标识,学习模型是为了推断出数据的一些内在结构。常见的应用场景包括关联规则的学习以及聚类等。常见算法包括Apriori算法和K-means算法。这类学习类型的目标不是让效用函数最大化,而是找到训练数据中的近似点。聚类常常能发现那些与假设匹配得相当好的直观分类,例如基于人口统计的聚合个体可能会在一个群体中形成一个富有的聚合,以及其他的贫穷的聚合。

3. 半监督学习

半监督学习(Semi-supervised Learning)是介于监督学习与无监督学习之间的一种机器学习方式,是模式识别和机器学习领域研究的重点问题。它主要考虑如何利用少量的标注样本和大量的未标注样本进行训练和分类的问题。半监督学习对于减少标注代价、提高学习机器性能具有非常重大的实际意义。半监督学习主要算法有基于概率的算法、在现有监督算法基础上进行修改的方法、直接依赖于聚类假设的方法等。在此学习方式下,输入数据部分被标识,其余部分没有被标识,这种学习模型可以用来进行预测,但是模型首先需要学习数据的内在结构以便合理地组织数据来进行预测。应用场景包括分类和回归;算法包括一些对常用监督式学习算法的延伸,这些算法首先试图对未标识数据进行建模,在此基础上再对标识的数据进行预测,如图论推理算法(Graph Inference)或者拉普拉斯支持向量机(Laplacian SVM)等。

半监督学习分类算法提出的时间比较短,还有许多方面没有更深入的研究。半监督学习从诞生以来,主要用于处理人工合成数据,无噪声干扰的样本数据是当前大部分半监督学习方法使用的数据,而在实际生活中用到的数据大部分不是无干扰的,通常都比较难以得到纯样本数据。

4. 强化学习

强化学习通过观察来学习动作的完成,每个动作都会对环境有所影响,学习对象根据观察到的周围环境的反馈来做出判断。在这种学习模式下,输入数据作为对模型的反馈,不像监督模型那样,输入数据仅仅是作为一个检查模型对错的方式。在强化学习下,输入数据直接反馈到模型,模型必须对此立刻做出调整。常见的应用场景包括动态系统以及机器人控制等。常见的算法包括 Q-Learning 以及时序差分学习(Temporal Difference Learning)。

12.2　MLlib 简介

MLlib 是 Spark 的机器学习库,旨在简化机器学习的工程实践工作,并方便扩展到更大规模。MLlib 由一些通用的学习算法和工具组成,包括分类、回归、聚类、协同过滤、降维等,同时还包括底层的优化原语和高层的管道 API,MLlib 库的组成部分如图 12-2 所示。

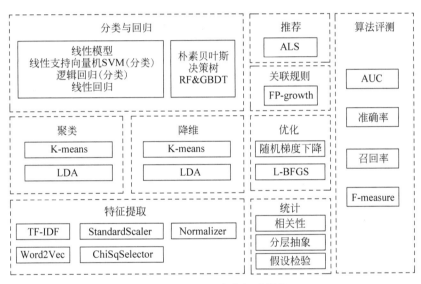

图 12-2　MLlib 库的组成部分

Spark 机器学习库从 1.2 版本以后被分为以下两个包。

(1) spark.mllib:包含基于 RDD 的原始机器学习算法 API。Spark MLlib 历史比较长,1.0 以前的版本已经包含了该机器学习包,提供的算法实现都是基于原始的 RDD。

(2) spark.ml:提供了基于 DataFrames 高层次的 API,可以用来构建机器学习管道(Pipeline)。ML Pipeline 弥补了原始 MLlib 库的不足,向用户提供了一个基于 DataFrame 的机器学习工作流式 API 套件。使用 ML Pipeline API 可以很方便地把数据处理、特征转换、正则化以及多个机器学习算法联合起来,构建一个单一完整的机器学习流水线。这种方式提供了更灵活的方法,更符合机器学习过程的特点,也更容易从其他语言进行迁移。Spark 官方推荐使用 spark.ml。如果新的算法能够适用于机器学习管道的概念,就应该将其放到 spark.ml 包中,如特征提取器和转换器。开发者需要注意的是,从 Spark 2.0 开始,基于 RDD 的 API 进入维护模式(即不增加任何新的特性),并计划于

Spark 3.0 版本时被移出 MLLib。因此，下面将以 ml 包为主进行介绍。

Spark MLlib 目前已经支持了主流的统计和机器学习算法。MLlib 目前支持 4 种常见的机器学习问题：分类、回归、聚类和协同过滤。

12.3 Spark MLlib 的数据类型

12.3.1 本地向量

1. 密集向量

密集向量是由 Double 类型的数组支持的，例如，向量(1.0,0.0,3.0)的密集向量表示格式为[1.0,0.0,3.0]。示例代码及运行结果如下：

```
scala> import org.apache.spark.mllib.linalg.{Vector, Vectors}
import org.apache.spark.mllib.linalg.{Vector, Vectors}
scala> val dv:Vector = Vectors.dense(1.0,0.0,3.0)
dv: org.apache.spark.mllib.linalg.Vector = [1.0,0.0,3.0]
```

2. 稀疏向量

稀疏向量是由两个并列的数组支持的，例如向量(1.0,0.0,3.0)的稀疏向量表示格式为(3,[0,2],[1.0,3.0])，其中 3 是向量(1.0,0.0,3.0)的长度，[0,2]是向量中非 0 维度的索引值，即向量索引 0 和 2 的位置为非 0 元素，[1.0,3.0]是按索引排列的数组元素值。示例代码及运行结果如下：

```
scala> import org.apache.spark.mllib.linalg.{Vector, Vectors}
import org.apache.spark.mllib.linalg.{Vector, Vectors}
scala> val sv1: Vector = Vectors.sparse(3, Array(0, 2), Array(1.0, 3.0))
sv1: org.apache.spark.mllib.linalg.Vector = (3,[0,2],[1.0,3.0])
```

12.3.2 标注点

标注点是一种带有标签的本地向量，通常用于监督学习算法中。MLlib 使用 Double 数据类型存储标签，因此可以在回归和分类中使用标记点。示例代码及运行结果如下：

```
scala> import org.apache.spark.mllib.regression.LabeledPoint
import org.apache.spark.mllib.regression.LabeledPoint
scala> val pos = LabeledPoint(1.0, Vectors.dense(1.0, 0.0, 3.0))
pos: org.apache.spark.mllib.regression.LabeledPoint = (1.0,[1.0,0.0,3.0])
```

12.3.3 本地矩阵

1. 密集矩阵

密集矩阵将所有元素的值存储在一个列优先的双精度数组中。示例代码及运行结果如下：

```
scala> import org.apache.spark.mllib.linalg.{Matrix, Matrices}
import org.apache.spark.mllib.linalg.{Matrix, Matrices}

scala> val dm: Matrix =Matrices.dense(3, 2, Array(1.0, 3.0, 5.0, 2.0, 4.0, 6.0))
dm: org.apache.spark.mllib.linalg.Matrix =
1.0  2.0
3.0  4.0
5.0  6.0
```

2. 稀疏矩阵

稀疏矩阵则将以列优先的非零元素压缩到稀疏列(CSC)格式中。示例如下。

创建一个3行2列的稀疏矩阵[[9.0,0.0],[0.0,8.0],[0.0,6.0]]。

第一个数组参数 Array(0,1,3)表示列指针,表示每一列非零元素的索引值。

第二个数组参数 Array(0,2,1)表示行索引,表示对应的非零元素是属于哪一行。

第三个数组 Array(9,6,8)是按列优先排序的所有非零元素,通过列指针和行索引即可判断每个元素所在的位置。

示例代码及运行结果如下:

```
scala> import org.apache.spark.mllib.linalg.{Matrix, Matrices}
import org.apache.spark.mllib.linalg.{Matrix, Matrices}

scala>  val sm: Matrix = Matrices.sparse(3, 2, Array(0, 1, 3), Array(0, 2, 1),
     |     Array(9, 6, 8))
sm: org.apache.spark.mllib.linalg.Matrix =
3 x 2 CSCMatrix
(0,0) 9.0
(2,1) 6.0
(1,1) 8.0
```

12.4 Spark MLlib 机器学习示例

12.4.1 特征抽取——TF-IDF

TF-IDF 是两个统计量的乘积,即词频(Term Frequency,TF)和逆向文档频率(Inverse Document Frequency,IDF)。它们各自有不同的计算方法。

TF 是一个文档(去除停用词之后)中某个词出现的次数。它用来度量词对文档的重要程度,TF 越大,该词在文档中就越重要。IDF 是逆向文档频率,是指文档集合中的总文档数除以含有该词的文档数,再取以 10 为底的对数。

TF-IDF 的主要思想是如果某个词或短语在一篇文章中出现的频率高(即 TF 高),并且在其他文章中很少出现(即 IDF 高),则认为此词或者短语具有很好的类别区分能力,适合用来分类。

具体实现如下。

(1)新建 MAVEN 项目,名称为 spark-mlllib,新建过程见第 9 章。在 pom.xml 文件中加入的依赖包如下:

```
<dependencies>
    <!-- 引入 Spark 相关的 JAR 包 -->
    <dependency>
        <groupId>org.apache.spark</groupId>
        <artifactId>spark-core_2.11</artifactId>
```

```xml
            <version>${spark.version}</version>
        </dependency>
        <dependency>
            <groupId>org.apache.spark</groupId>
            <artifactId>spark-sql_2.11</artifactId>
            <version>${spark.version}</version>
        </dependency>
        <dependency>
            <groupId>org.apache.spark</groupId>
            <artifactId>spark-streaming_2.11</artifactId>
            <version>${spark.version}</version>
        </dependency>
        <dependency>
            <groupId>org.apache.spark</groupId>
            <artifactId>spark-mllib_2.11</artifactId>
            <version>${spark.version}</version>
        </dependency>
</dependencies>
```

（2）数据准备。新建一个文本文件，包含 4 行数据，内容如下：

```
hello mllib hello spark
goodBye spark
hello spark
goodBye spark
```

（3）新建 Scala 类，其功能是计算单词的 TF-IDF。相关代码及说明如下：

```scala
import org.apache.spark.mllib.feature.{HashingTF, IDF}
import org.apache.spark.mllib.linalg
import org.apache.spark.rdd.RDD
import org.apache.spark.{SparkConf, SparkContext}
object TF-IDF{
  def main(args: Array[String]){
    //创建环境变量
    val conf = new SparkConf()
    //设置本地化处理
    .setMaster("local")
    //设定名称
    .setAppName("TF-IDF") //设定名称
    val sc = new SparkContext(conf)
    //设置日志级别
    sc.setLogLevel("error")
    //读取数据并将句子分割成单词
    val documents = sc.textFile("a.txt")
    .map(_.split(" ").toSeq)
    println("分词的结果为:")
    documents.foreach(println)
    //创建 TF 计算实例
    val hashingTF = new HashingTF()
```

```
    //计算文档 TF 值
    val tf = hashingTF.transform(documents).cache()
    println("计算单词出现的次数结果为:")
    tf.foreach(println)
    //创建 IDF 实例并计算
    val idf = new IDF().fit(tf)
    //计算 TF - IDF 词频
    val tf_idfRDD: RDD[linalg.Vector] = idf.transform(tf)
    println("计算 TF - IDF 值:")
    tf_idfRDD.foreach(println)
  }
}
```

（4）运行程序。TF-IDF 示例程序运行结果如图 12-3 所示。

```
分词的结果为:
WrappedArray(hello, mllib, hello, spark)
WrappedArray(goodBye, spark)
WrappedArray(hello, spark)
WrappedArray(goodBye, spark)
计算单词出现的次数结果为:
(1048576,[165160,496801,690825],[2.0,1.0,1.0])
(1048576,[5033,496801],[1.0,1.0])
(1048576,[165160,496801],[1.0,1.0])
(1048576,[5033,496801],[1.0,1.0])
计算TF-IDF值:
(1048576,[165160,496801,690825],[1.0216512475319814,0.0,0.9162907318741551])
(1048576,[5033,496801],[0.5108256237659907,0.0])
(1048576,[165160,496801],[0.5108256237659907,0.0])
(1048576,[5033,496801],[0.5108256237659907,0.0])
```

图 12-3　TF-IDF 示例程序运行结果

计算单词出现的次数运行结果的第一行"1048576,[165160,496801,690825],[2.0,1.0,1.0]"中，1048576 代表哈希表的桶数，"[165160,496801,690825]"分别代表"hello""spark""mllib"的哈希值，"[2.0,1.0,1.0]"为对应单词的出现次数。

计算 TF-IDF 值运行结果第一行"[1.0216512475319814,0.0,0.9162907318741551]"分别代表"hello""spark""mllib"的 TF-IDF 值。注意 spark 这个单词，计算出来的 TF-IDF 值是 0，因为这个单词在所有行中都出现了，它的 IDF 逆向文档频率为 0，表明这个单词的没有对文档的区分能力。

12.4.2　分类与回归——线性回归

此例通过工具类 MLUtils 加载 LIBSVM 格式样本文件，每一行的第一个是真实值 y，有 10 个特征值 x，用 1:double,2:double 分别标注，即建立需求函数：

$$y = a_1 x_1 + a_2 x_{2} + a_3 x_3 + a_4 x_4 + \cdots + a_{10} x_{10}$$

通过样本数据和梯度下降训练模型，找到 10 个比较合理的参数值（$a_1 \sim a_{10}$）。

相应代码和说明如下：

```scala
import org.apache.spark.mllib.regression.{LabeledPoint, LinearRegressionWithSGD}
import org.apache.spark.mllib.util.MLUtils
import org.apache.spark.rdd.RDD
import org.apache.spark.{SparkConf, SparkContext}
object LinearRegressionDemo{
  def main(args: Array[String]): Unit = {
    //创建 SparkContext
    val conf = new SparkConf().setMaster("local[4]").setAppName("LinearRegression")
    val sc = new SparkContext(conf)
    sc.setLogLevel("error")
    //加载数据样本
    val path = "data1.txt"
    //通过提供的工具类加载样本文件,每一行的第一个是 y 值,有 10 个特征值 x,用 1:double,
    //2:double 分别标注
    val data: RDD[LabeledPoint] = MLUtils.loadLibSVMFile(sc, path).cache()
    //迭代次数
    val numIterations = 100
    //梯度下降步长
    val stepSize = 0.00000001
    //训练模型
    val model = LinearRegressionWithSGD.train(data,numIterations,stepSize)
    //模型评估
    val valuesAndPreds = data.Map{ point =>
    //根据模型预测 Label 值
    val prediction = model.predict(point.features)
    println(s"【真实值】: ${point.label};【预测值】: ${prediction}")
      (point.label,prediction)
    }
    //打印模型参数
    println("【参数值】: " + model.weights)
    //求均方误差
    val MSE = valuesAndPreds.Map{ case(v, p) => math.pow((v - p) , 2) }.mean()
    println("训练模型的均方误差为 = " + MSE)
    sc.stop()
  }
}
```

线性回归示例程序运行结果如图 12-4 所示。

图 12-4　线性回归示例程序运行结果

12.4.3　分类与回归——逻辑回归

此例通过建立随机梯度下降的回归模型预测胃癌是否转移,数据特征说明如下:

y：胃癌转移情况（有转移 $y=1$；无转移 $y=0$）。

x_1：确诊时患者的年龄（岁）。

x_2：肾细胞癌血管内皮生长因子（VEGF），其阳性表述由低到高共 3 个等级。

x_3：肾细胞癌组织内微血管数（MVC）。

x_4：肾癌细胞核组织学分级，由低到高共 4 级。

x_5：肾癌细胞分期，由低到高共 4 期。

代码及说明如下：

```
import org.apache.spark.{SparkConf, SparkContext}
import rg.apache.spark.mllib.classification.{LogisticRegressionWithLBFGS, LogisticRegressionWithSGD}
import org.apache.spark.mllib.evaluation.MulticlassMetrics
import org.apache.spark.mllib.linalg.Vectors
import org.apache.spark.mllib.regression.LabeledPoint
import org.apache.spark.mllib.util.MLUtils
object LogisticRegressionDemo{
  def main(args: Array[String]): Unit = {
    //建立 Spark 环境
    val conf = new SparkConf().setAppName("logisticRegression").setMaster("local")
    val sc = new SparkContext(conf)
    sc.setLogLevel("error")
    //通过 MLUtils 工具类读取 LIBSVM 格式数据集
    val data = MLUtils.loadLibSVMFile(sc, "wa.txt")
    //测试集和训练集按 2:8 的比例分
    val Array(traning,test) = data.randomSplit(Array(0.8, 0.2) , seed = 1L)
    println(traning.count,test.count)
    traning.foreach(println)
    //建立 LogisticRegressionWithLBFGS 对象，设置分类数 2，传入训练集开始训练，返回训练后的模型
    val model = new LogisticRegressionWithLBFGS()
      .setNumClasses(2)
      .run(traning)
    //使用训练后的模型对测试集进行测试，同时打印标签和测试结果
    val labelAndPreds = test.Map{ point =>
      val prediction = model.predict(point.features)
      (point.label,prediction)
    }
    labelAndPreds.foreach(println)
    println("推荐" + model.weights)
    val trainErr = labelAndPreds.filter( r => r._1 != r._2).count.toDouble / test.count
    println("容错率为 trainErr: " + trainErr)
    val predictionAndLabels = test.Map{
    //计算测试值
    case LabeledPoint(label,features) =>
        val prediction = model.predict(features)  (prediction,label)    }
    //创建验证类
    val metrics = new MulticlassMetrics(predictionAndLabels)
    val precision = metrics.precision
    //计算验证值
    println("Precision = " + precision)
```

```
//利用模型进行新患者预测
val patient = Vectors.dense(Array(20, 1, 0.0, 1, 1))
val d = model.predict(patient)
print("预测的结果为:" + d)
//计算患者可能性
if(d = = 1){
    println("患者的胃癌有概率转移.")
} else {
    println("患者的胃癌没有概率转移.")
}
```

逻辑回归示例程序运行结果如图 12-5 所示。

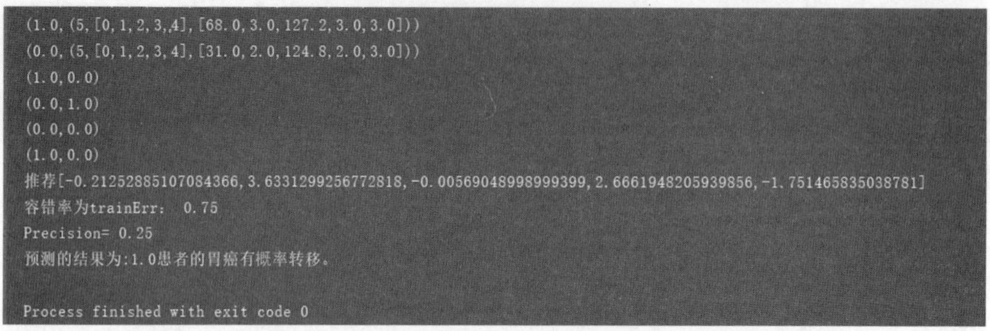

```
(1.0,(5,[0,1,2,3,4],[68.0,3.0,127.2,3.0,3.0]))
(0.0,(5,[0,1,2,3,4],[31.0,2.0,124.8,2.0,3.0]))
(1.0,0.0)
(0.0,1.0)
(0.0,0.0)
(1.0,0.0)
推荐[-0.21252885107084366,3.6331299256772818,-0.00569048998999399,2.6661948205939856,-1.751465835038781]
容错率为trainErr: 0.75
Precision= 0.25
预测的结果为:1.0患者的胃癌有概率转移.

Process finished with exit code 0
```

图 12-5　逻辑回归示例程序运行结果

12.4.4　协同过滤——电影推荐

协同过滤是利用大量已有的用户偏好来估计用户对其未接触过的物品的喜好程度。在协同过滤算法中有着两个分支,分别是基于群体用户的协同过滤(UserCF)和基于物品的协同过滤(ItemCF)。

在电影推荐系统中,通常分为针对用户推荐电影和针对电影推荐用户两种方式。若采用基于用户的推荐模型,则会利用相似用户的评级来计算对某个用户的推荐。若采用基于电影的推荐模型,则会依靠用户接触过的电影与候选电影之间的相似度来获得推荐。

在 Spark MLlib 中实现了交替最小二乘(ALS)算法,它是机器学习的协同过滤式推荐算法。机器学习的协同过滤式推荐算法是通过观察所有用户给产品的评分来推断每个用户的喜好,并向用户推荐合适的产品。具体步骤如下。

(1) 下载数据集,网址为 http://files.grouplens.org/datasets/movielens/ml-1m.zip。
(2) 建立 Scala 类 MovieRecomment,相应代码和说明如下:

```
import java.util.Random
import org.apache.spark.SparkConf
import org.apache.spark.SparkContext
import org.apache.spark.rdd._
import org.apache.spark.mllib.recommendation.{ALS,Rating,MatrixFactorizationModel}
```

```scala
object MovieRecomment{
    def main(args: Array[String]): Unit = {
    //建立 Spark 环境
    val conf = new SparkConf().setAppName("MovieLensALS").setMaster("local[4]")
    val sc = new SparkContext(conf)
    sc.setLogLevel("error")
    //装载样本评分数据,其中最后一列 Timestamp 取除以 10 的余数作为 key,Rating 为值
    val ratings = sc.textFile("ratings.dat").Map{
      line =>
        val fields = line.split("::")
        //时间戳、用户编号、电影编号、评分
        (fields(3).toLong % 10,Rating(fields(0).toInt, fields(1).toInt, fields(2).toDouble))
            }
    ratings.take(10).foreach { println }
    //装载电影目录对照表(电影 ID->电影标题)
    val movies = sc.textFile("movies.dat").Map{ line =>
      val fields = line.split("::")
        (fields(0).toInt, fields(1))
    }.collect.toMap
    //记录样本评分数、用户数、电影数
    val numRatings = ratings.count
    val numUsers = ratings.Map(_._2.user).distinct.count
    val numMovies = ratings.Map(_._2.product).distinct.count
    println("从" + numRatings + "记录中" + "分析了" + numUsers + "的人观看了" +
numMovies + "部电影")
    //提取一个得到最多评分的电影子集(电影编号),以便进行评分启发
    //矩阵最为密集的部分
    val mostRatedMovieIds = ratings.Map(_._2.product)
      .countByValue()
      .toSeq
      .sortBy(-_._2)
      .take(50)           //取前 50 个
      .Map(_._1)          //获取他们的 id
    val random = new Random(0)
    //从目前最火的电影中随机获取 10 部电影
    val selectedMovies = mostRatedMovieIds.filter(
      x => random.nextDouble() < 0.2).Map(x => (x, movies(x))).toSeq
    //调用函数 elicitateRatings(),引导或者启发评论,让用户打分
    val myRatings = elicitateRatings(selectedMovies)
    val myRatingsRDD = sc.parallelize(myRatings)
    //将评分数据分成训练集 60%、验证集 20%、测试集 20%
    val numPartitions = 20
    //训练集
    val training = ratings.filter(x => x._1 < 6).values
      .union(myRatingsRDD).repartition(numPartitions)
      .persist
    //验证集
    val validation = ratings.filter(x => x._1 >= 6 && x._1 < 8).values
      .repartition(numPartitions).persist
```

```scala
//测试集
val test = ratings.filter(x => x._1 >= 8).values.persist
val numTraining = training.count
val numValidation = validation.count
val numTest = test.count
println("训练集数量:" + numTraining + ",验证集数量:" + numValidation + ", 测试集数量:" + numTest)
//训练模型,并且在验证集上评估模型
//特征向量纬度
val ranks = List(8, 12)
val lambdas = List(0.1, 10.0)
//循环次数
val numIters = List(10, 20)
var bestModel: Option[MatrixFactorizationModel] = None
var bestValidationRmse = Double.MaxValue
var bestRank = 0
var bestLambda = -1.0
var bestNumIter = -1
for (rank <- ranks; lambda <- lambdas; numIter <- numIters){
    val model = ALS.train(training, rank, numIter, lambda)
    val validationRmse = computeRmse(model, validation, numValidation)
    println("RMSE (validation) = " + validationRmse + "for the model trained with rand = " + rank + ", lambda = " + lambda + ", and numIter = " + numIter + ".")
    if (validationRmse < bestValidationRmse){
    bestModel = Some(model)
    bestValidationRmse = validationRmse
    bestRank = rank
    bestLambda = lambda
    bestNumIter = numIter
    }
}
//在测试集中获得最佳模型
val testRmse = computeRmse(bestModel.get,tes, numTest)
println("The best model was trained with rank = " + bestRank + " and lambda = " + bestLambda + ", and numIter = " + bestNumIter + ", and itsRMSE on the test set is" + testRmse + ".")
//产生个性化推荐
val myRateMoviesIds = myRatings.Map(_.product).toSet
val candidates = sc.parallelize(movies.keys.filter(!myRateMoviesIds.contains(_)).toSeq)
val recommendations = bestModel.get.predict(candidates.Map((0, _)))
    .collect()
    .sortBy((-_.rating))
    .take(10)
var i = 1
println("以下电影推荐给你")
recommendations.foreach{ r =>
    println("%2d".format(i) + ":" + movies(r.product))
    i += 1
}
```

```
}
/** 计算均方根误差 RMSE */
def computeRmse(model: MatrixFactorizationModel,data: RDD[Rating], n: Long) = {
  val predictions: RDD[Rating] = model.predict(data.Map(x => (x.user,x.product)))
  val predictionsAndRatings = predictions.Map(x => ((x.user,x.product), x.rating))
    .join(data.Map(x => ((x.user,x.product),x.rating)))
    .values
  math.sqrt(predictionsAndRatings.Map(x => (x._1 - x._2) * (x._1 - x._2)).Reduce(_ + _) / n)
}
/** 从命令行获取用户电影评级. */
def elicitateRatings(movies: Seq[(Int,String)]) = {
  val prompt = "给以下电影评分(1~5分)"
  println(prompt)
  val ratings = movies.flatMap{ x =>
   var rating: Option[Rating] = None
    var valid = false
    while (!valid){
      print(x._2 + ": ")
      try {
        val r = Console.readInt
        if (r < 0 || r > 5){
          println(prompt)
        } else {
          valid = true
          if (r > 0){
            rating = Some(Rating(0, x._1,r))
          }
        }
      } catch {
        case e: Exception => println(prompt)
      }
    }
    rating match{
      case Some(r) => Iterator(r)
      case None => Iterator.empty
    }
  }
  if (ratings.isEmpty){
    error("No rating provided!")
  } else {
    ratings
  }
}


}
}
```

（3）结合用户输入的评分，程序给用户推荐了 10 个可能感兴趣的电影。协同过滤示例程序运行结果如图 12-6 所示。

```
The best model was trained with rank=12 and lambda =0.1, and numIter =20, and itsRMSE on the test set is0.8687552593996356.
以下电影推荐给你
  1:Bewegte Mann, Der (1994)
  2:Sanjuro (1962)
  3:For All Mankind (1989)
  4:Leather Jacket Love Story (1997)
  5:Man of the Century (1999)
  6:Shawshank Redemption, The (1994)
  7:Usual Suspects, The (1995)
  8:Bandits (1997)
  9:American Beauty (1999)
 10:Schindler's List (1993)

Process finished with exit code 0
```

图 12-6　协同过滤示例程序运行结果

12.5　本章小结

本章介绍了机器学习的基本概念和分类，重点介绍了 Spark MLlib 目前包含的算法和组件，通过 4 个具体例子展示了利用 Spark MLlib 进行机器学习的方法和步骤，这些例子需要读者掌握机器学习的基本知识和算法的基本原理，这些知识在本书中没有进行详细的介绍，读者可以查看机器学习的相关书籍，以加深对机器学习的进一步理解。

第13章

实战案例——分布式优惠券后台应用系统

13.1 系统简介

分布式优惠券后台应用系统服务于两类用户群体：一类是商户，商户可以根据自己的实际情况进行优惠券投放；另一类是平台消费用户，用户可以去领取商户发放的优惠券。

13.2 整体架构

分布式优惠券后台应用系统采用 SpringBoot 作为主体开发框架，使用 Kafka 消息队列实现优惠券从商户到用户的传递，MySQL 存储商户信息，HBase 存储用户信息、优惠券信息等，Redis 保存优惠券的缓存信息。系统整体架构如图 13-1 所示。

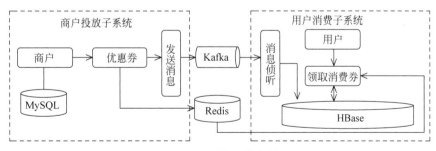

图 13-1　系统整体架构

对于商户投放子系统，商户注册生成对应的商户实体信息，并保存到 MySQL 数据库中(Merchants)，商户可以投放自己商家的优惠券(PassTemplate)，且优惠券有自己的

Token 存放于 Redis 中，投放的优惠券信息将由 Kafka 向用户消费子系统发送，而商户投放的优惠券并不在 MySQL 中进行存储，而是在商户投放子系统中放送消息给 Kafka，用户消费子系统通过侦听 Kafka 消息获得 Kafka 分布式信息并存储到 HBase 中。用户通过读取 Redis 中的优惠券信息领取消费券，并将自己领取到的优惠券信息存放在 HBase 中。

13.3 表结构设计

（1）MySQL 的表结构设计。在 MySQL 中建立一个商户表，存放商户基本信息。Merchants（商户表）如表 13-1 所示。

表 13-1 Merchants（商户表）

字段名	类型	大小	是否主键	说明
id	int	10	是	商户 ID
name	varchar	64		商户名称
logo_url	varchar	256		商户 logo
business_license_url	varchar	256		商户营业执照
phone	varchar	64		商户联系电话
address	varchar	64		商户地址
is_audit	boolean			是否通过审核

（2）HBase 表结构设计。在 HBase 中建立三个表，分别是消费用户表、优惠券表和优惠券领取表。pb:user（消费用户表）如表 13-2 所示。pb:passtemplate（优惠券表）如表 13-3 所示。pb:pass（优惠券领取表）如表 13-4 所示。

表 13-2 pb:user（消费用户表）

基本信息列族（b）	
name	用户名
age	用户年龄
sex	用户性别
额外信息列族（o）	
phone	电话号码
address	住址

表 13-3 pb:passtemplate（优惠券表）

基本信息列族（b）	
id	商户 id
title	优惠券标题
summary	优惠券摘要信息
desc	优惠券详细信息
has_token	优惠券是否有 Token
background	优惠券背景色

续表

约束信息列族（c）	
limit	最大个数限制
start	优惠券开始时间
end	优惠券结束时间

表 13-4　pb：pass（优惠券领取表）

信息列族（i）	
user_id	用户 id
template_id	优惠券 id
token	优惠券识别码
assigned_date	领取日期
con_date	消费日期

13.4　系统实现

13.4.1　商户投放子系统

（1）新建 MAVEN 工程，引入相关依赖包。pom.xml 文件内容如下：

```xml
<?xml version="1.0" encoding="UTF-8"?>
<project xmlns=http://maven.apache.org/POM/4.0.0
  xmlns:xsi="http://www.w3.org/2001/XMLSchema-instance"
  xsi:schemaLocation="http://maven.apache.org/POM/4.0.0
  http://maven.apache.org/xsd/maven-4.0.0.xsd">
    <modelVersion>4.0.0</modelVersion>
    <groupId>com.coupon.passbook</groupId>
    <artifactId>merchants</artifactId>
    <version>0.0.1-SNAPSHOT</version>
    <name>merchants</name>
    <description>passbook of merchants</description>
    <parent>
        <groupId>org.springframework.boot</groupId>
        <artifactId>spring-boot-starter-parent</artifactId>
        <version>1.5.3.RELEASE</version>
        <relativePath/> <!-- lookup parent from repository -->
    </parent>
    <dependencies>
        <dependency>
            <groupId>mysql</groupId>
            <artifactId>mysql-connector-java</artifactId>
            <version>5.1.41</version>
        </dependency>
        <dependency>
            <groupId>org.springframework.boot</groupId>
```

```xml
        <artifactId>spring-boot-starter-data-jpa</artifactId>
    </dependency>
    <dependency>
        <groupId>org.springframework.boot</groupId>
        <artifactId>spring-boot-starter-web</artifactId>
    </dependency>
    <dependency>
        <groupId>org.springframework.boot</groupId>
        <artifactId>spring-boot-starter-jdbc</artifactId>
    </dependency>
    <dependency>
        <groupId>org.yaml</groupId>
        <artifactId>snakeyaml</artifactId>
        <version>1.10</version>
    </dependency>
    <dependency>
        <groupId>org.springframework.boot</groupId>
        <artifactId>spring-boot-configuration-processor</artifactId>
        <optional>true</optional>
    </dependency>
    <dependency>
        <groupId>org.projectlombok</groupId>
        <artifactId>lombok</artifactId>
    </dependency>
    <dependency>
        <groupId>org.springframework.boot</groupId>
        <artifactId>spring-boot-starter-test</artifactId>
        <scope>test</scope>
    </dependency>
    <dependency>
        <groupId>com.alibaba</groupId>
        <artifactId>fastjson</artifactId>
        <version>1.2.31</version>
    </dependency>
    <dependency>
        <groupId>org.springframework.boot</groupId>
        <artifactId>spring-boot-autoconfigure</artifactId>
    </dependency>
    <dependency>
        <groupId>org.springframework.kafka</groupId>
        <artifactId>spring-kafka</artifactId>
        <version>1.1.1.RELEASE</version>
    </dependency>
    <dependency>
        <groupId>commons-lang</groupId>
        <artifactId>commons-lang</artifactId>
        <version>2.2</version>
    </dependency>
</dependencies>
```

```xml
<build>
    <finalName>Merchants</finalName>
    <plugins>
        <plugin>
            <groupId>org.springframework.boot</groupId>
            <artifactId>spring-boot-maven-plugin</artifactId>
        </plugin>
    </plugins>
</build>
</project>
```

（2）修改 application.yml 文件，文件中主要包括 MySQL 数据库和 Kafka 连接的相关配置，内容如下：

```yaml
spring:
  application:
    name: Merchants
  datasource:
    url: jdbc:mysql://127.0.0.1:3306/passbook?autoReconnect=true
    username: root
    password: 123456
  kafka:
    bootstrap-servers: dc3-data:9092
    consumer:
      group-id: passbook
    listener:
      concurrency: 4
server:
  port: 9527
logging:
  level: debug
  file: merchants.log
```

（3）建立各个类包存放路径。其中，constant 是系统的常量信息的存放路径，包括 Kafka Topic 名称、错误代码、优惠券颜色代码等信息；controller 是控制层实现类的存放路径；dao 是实现数据库存取的实现类的存放路径；entity 是实体类的存放路径；security 是安全访问控制包的存放路径；service 是业务层实现类包的存放路径；vo 是验证请求的有效性类包的存放路径。商户投放子系统类包存放路径如图 13-2 所示。

数据库访问层采用 Spring Data JPA。Spring Data JPA 是 Spring 基于 ORM 框架、JPA 规范的基础封装的一套 JPA 应用框架，可使开发者用极简的代码即可实现对数据的访问和操作。它提供了包括增加、删除、修改、查询等在内的常用功能，易于扩展，可以极大提高开发效率。

（4）核心代码实现。

商户子系统类包调用关系如图 13-3 所示。

图 13-2 商户投放子系统类包存放路径

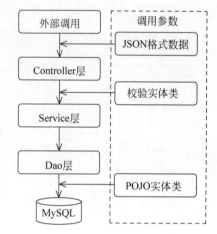

图 13-3 商户子系统类包调用关系

① Controller 层主要实现外部 post()和 get()调用的响应,代码实现如下:

```
package com.coupon.passbook.controller;
import com.alibaba.fastjson.JSON;
import com.coupon.passbook.service.IMerchantsServ;
import com.coupon.passbook.vo.CreateMerchantsRequest;
import com.coupon.passbook.vo.PassTemplate;
import com.coupon.passbook.vo.Response;
import lombok.extern.slf4j.Slf4j;
import org.springframework.beans.factory.annotation.Autowired;
import org.springframework.web.bind.annotation.GetMapping;
import org.springframework.web.bind.annotation.PathVariable;
import org.springframework.web.bind.annotation.PostMapping;
import org.springframework.web.bind.annotation.RequestBody;
import org.springframework.web.bind.annotation.RequestMapping;
import org.springframework.web.bind.annotation.ResponseBody;
import org.springframework.web.bind.annotation.RestController;


/**
 * <h1>商户服务 Controller</h1>
 * Created by hhao.
 */
@Slf4j
@RestController
@RequestMapping("/merchants")
```

```java
public class MerchantsCtl{
    /** 商户服务接口 */
    private final IMerchantsServ merchantsServ;
    @Autowired
    public MerchantsCtl(IMerchantsServ merchantsServ){
        this.merchantsServ = merchantsServ;
    }
    //创造商户
    @ResponseBody
    @PostMapping("/create")
        public Response createMerchants(@RequestBody CreateMerchantsRequest request)
    {
        log.info("CreateMerchants: {}",JSON.toJSONString(request));
        return merchantsServ.createMerchants(request);
    }
    //按商户 id 查询商户信息
    @ResponseBody
    @GetMapping("/{id}")
    public Response buildMerchantsInfo(@PathVariable Integer id){
        log.info("BuildMerchantsInfo: {}", id);
        return merchantsServ.buildMerchantsInfoById(id);
    }
    /** 商户发放优惠券,具体格式为:
     * DropPassTemplates:{"background":1,"desc":"详情:大数据优惠券 second",
     * "end":1528202373202,"hasToken":false,"id":9,"limit":1000,
     * "start":1527338373202,"summary":"简介:大数据优惠券","title":"title:大数据
       优惠券"}
     * */
    @ResponseBody
    @PostMapping("/drop")
        public Response dropPassTemplate(@RequestBody PassTemplate passTemplate){
        log.info("DropPassTemplate: {}", passTemplate);
        return merchantsServ.dropPassTemplate(passTemplate);
    }
}
```

② Service 层的主要功能是实现对表的插入操作和向 Kafka 发送消息,发送消息使用了 org.springframework.kafka.core.KafkaTemplate 下的 KafkaTemplate 客户端向 Kafka 发送优惠券消息,代码如下:

```java
package com.coupon.passbook.service.impl;
import com.alibaba.fastjson.JSON;
import com.coupon.passbook.constant.Constants;
import com.coupon.passbook.constant.ErrorCode;
import com.coupon.passbook.service.IMerchantsServ;
import com.coupon.passbook.vo.CreateMerchantsRequest;
import com.coupon.passbook.vo.CreateMerchantsResponse;
import com.coupon.passbook.vo.Response;
import com.coupon.passbook.dao.MerchantsDao;
```

```java
import com.coupon.passbook.entity.Merchants;
import com.coupon.passbook.vo.PassTemplate;
import lombok.extern.slf4j.Slf4j;
import org.springframework.beans.factory.annotation.Autowired;
import org.springframework.kafka.core.KafkaTemplate;
import org.springframework.stereotype.Service;
import org.springframework.transaction.annotation.Transactional;
/**
 * <h1>商户服务接口实现</h1>
 * Created by hhao.
 */
@Slf4j
@Service
public class MerchantsServImpl implements IMerchantsServ{
    /** Merchants 数据库接口 */
    private final MerchantsDao merchantsDao;
    /** Kafka 客户端 */
    private final KafkaTemplate<String,String> kafkaTemplate;
    @Autowired
    public MerchantsServImpl(MerchantsDao merchantsDao,
    KafkaTemplate<String,String> kafkaTemplate) {
        this.merchantsDao = merchantsDao;
        this.kafkaTemplate = kafkaTemplate;
    }
    @Override
    @Transactional
    //创建商户
    public Response createMerchants(CreateMerchantsRequest request){
      Response response = new Response();
            CreateMerchantsResponse merchantsResponse = new
      CreateMerchantsResponse();
      ErrorCode errorCode = request.validate(merchantsDao);
      //没有找到数据的处理
      if (errorCode != ErrorCode.SUCCESS){
          merchantsResponse.setId(-1);
          response.setErrorCode(errorCode.getCode());
          response.setErrorMsg(errorCode.getDesc());
      } else {


      merchantsResponse.setId(merchantsDao.save(request.toMerchants()).getId());
      }
      response.setData(merchantsResponse);
      return response;
    }
        //以商户 id 查询商户
        @Override
        public Response buildMerchantsInfoById(Integer id){
            Response response = new Response();
```

```java
            Merchants merchants = merchantsDao.findById(id);
            //没有找到商户数据的处理
            if (null == merchants){
            response.setErrorCode(ErrorCode.MERCHANTS_NOT_EXIST.getCode());

             response.setErrorMsg(ErrorCode.MERCHANTS_NOT_EXIST.getDesc());
            }
            response.setData(merchants);
            return response;
    }
        //商户发放优惠券,向 Kafka 发送消息
    @Override
    public Response dropPassTemplate(PassTemplate template){
        Response response = new Response();
        ErrorCode errorCode = template.validate(merchantsDao);
//没有此商户的异常处理
        if (errorCode != ErrorCode.SUCCESS){
            response.setErrorCode(errorCode.getCode());
            response.setErrorMsg(errorCode.getDesc());
        }
//系统有此商户,发入优惠券,向 Kafka 发送消息
      else{
            String passTemplate = JSON.toJSONString(template);
            log.info("passTemplate = " + passTemplate + "TEMPLATE_TOPIC = = " +
            Constants.TEMPLATE_TOPIC);
            kafkaTemplate.send(
                    Constants.TEMPLATE_TOPIC,
                    Constants.TEMPLATE_TOPIC,
                    passTemplate
            );
            log.info("DropPassTemplates: {}",passTemplate);
        }
        return response;
    }
}
```

③ Dao 层实现与数据库进行交互,代码如下:

```java
package com.coupon.passbook.dao;
import com.coupon.passbook.entity.Merchants;
import org.springframework.data.jpa.repository.JpaRepository;
/**
 * <h1>Merchants Dao 接口</h1>
 * Created by hhao.
 */
public interface MerchantsDao extends JpaRepository<Merchants, Integer> {
    /**
     * <h2>根据 id 获取商户对象</h2>
     * @param id 商户 id
     * @return {@link Merchants}
```

```
     * */
    Merchants findById(Integer id);
    /**
     * <h2>根据商户名称获取商户对象</h2>
     * @param name 商户名称
     * @return {@link Merchants}
     * */
    Merchants findByName(String name);
}
```

13.4.2 用户消费子系统

(1) 新建 MAVEN 工程,引入相关依赖包。pom.xml 文件内容如下:

```
<?xml version = "1.0" encoding = "UTF-8"?>
    <project xmlns = http://maven.apache.org/POM/4.0.0
xmlns:xsi = "http://www.w3.org/2001/XMLSchema-instance"
    xsi:schemaLocation = "http://maven.apache.org/POM/4.0.0
http://maven.apache.org/xsd/maven-4.0.0.xsd">
        <modelVersion>4.0.0</modelVersion>
        <groupId>com.coupon.passbook</groupId>
        <artifactId>passbook</artifactId>
        <version>0.0.1-SNAPSHOT</version>
        <packaging>jar</packaging>
        <name>passbook</name>
        <description>passbook</description>
        <parent>
            <groupId>org.springframework.boot</groupId>
            <artifactId>spring-boot-starter-parent</artifactId>
            <version>1.5.3.RELEASE</version>
            <relativePath/> <!-- lookup parent from repository -->
        </parent>
        <dependencies>
            <dependency>
                <groupId>mysql</groupId>
                <artifactId>mysql-connector-java</artifactId>
                <version>5.1.41</version>
            </dependency>
            <dependency>
                <groupId>org.springframework.boot</groupId>
                <artifactId>spring-boot-starter-data-jpa</artifactId>
            </dependency>
            <dependency>
                <groupId>org.springframework.boot</groupId>
                <artifactId>spring-boot-starter-web</artifactId>
            </dependency>
            <dependency>
                <groupId>org.springframework.boot</groupId>
                <artifactId>spring-boot-starter-jdbc</artifactId>
```

```xml
</dependency>
<dependency>
    <groupId>org.springframework.boot</groupId>
    <artifactId>spring-boot-starter-data-redis</artifactId>
</dependency>
<dependency>
    <groupId>org.yaml</groupId>
    <artifactId>snakeyaml</artifactId>
    <version>1.10</version>
</dependency>
<dependency>
    <groupId>com.spring4all</groupId>
    <artifactId>spring-boot-starter-hbase</artifactId>
    <version>1.0.0.RELEASE</version>
</dependency>
<dependency>
    <groupId>org.springframework.boot</groupId>
    <artifactId>spring-boot-configuration-processor</artifactId>
    <optional>true</optional>
</dependency>
<dependency>
    <groupId>org.projectlombok</groupId>
    <artifactId>lombok</artifactId>
</dependency>
<dependency>
    <groupId>org.springframework.boot</groupId>
    <artifactId>spring-boot-starter-test</artifactId>
    <scope>test</scope>
</dependency>
<dependency>
    <groupId>com.alibaba</groupId>
    <artifactId>fastjson</artifactId>
    <version>1.2.31</version>
</dependency>
<dependency>
    <groupId>org.apache.commons</groupId>
    <artifactId>commons-lang3</artifactId>
    <version>3.0</version>
</dependency>
<dependency>
    <groupId>org.springframework.boot</groupId>
    <artifactId>spring-boot-autoconfigure</artifactId>
</dependency>
<dependency>
    <groupId>org.springframework.kafka</groupId>
    <artifactId>spring-kafka</artifactId>
    <version>1.1.1.RELEASE</version>
</dependency>
<dependency>
```

```xml
            <groupId>org.springframework.boot</groupId>
            <artifactId>spring-boot-starter-thymeleaf</artifactId>
        </dependency>
        <dependency>
            <groupId>org.springframework.boot</groupId>
            <artifactId>spring-boot-devtools</artifactId>
            <optional>true</optional>
        </dependency>
    </dependencies>
    <build>
        <finalName>UserPassbook</finalName>
        <plugins>
            <plugin>
                <groupId>org.springframework.boot</groupId>
                <artifactId>spring-boot-maven-plugin</artifactId>
                <configuration>
                    <fork>true</fork>
                </configuration>
            </plugin>
            <plugin>
                <groupId>org.apache.maven.plugins</groupId>
                <artifactId>maven-compiler-plugin</artifactId>
                <configuration>
                    <source>8</source>
                    <target>8</target>
                </configuration>
            </plugin>
        </plugins>
    </build>
</project>
```

（2）修改 application.yml 文件，包括 MySQL、HBase、Redis 数据库和 Kafka 连接的相关配置，内容如下：

```yml
spring:
  application:
    name: Passbook
  datasource:
    url: jdbc:mysql://127.0.0.1:3306/passbook?autoReconnect=true
    username: root
    password: 123456
  kafka:
    bootstrap-servers: dc3-data:9092
    consumer:
      group-id: passbook
    listener:
      concurrency: 4
  data:
    hbase:
      quorum: dc3.com:2181
      rootDir: file:/passbook/tmp/hbase/root
```

```
      nodeParent: /hbase
  redis:
    host: 127.0.0.1
    port: 6379
    password: 123456
server:
  port: 9528
logging:
  file: assbook.log
  level: debug
```

（3）建立各个实现类包的存放路径。其中，advice 是全局异常处理类的存放路径；constant 是系统的常量信息存放路径，包括 Kafka Topic 名称、错误代码、HBase 中的表名和列族信息等；controller 是控制层的实现类的存放路径；dao 是实现 MySQL 数据库存取操作的实现类的存放路径；entity 是实体类的存放路径；log 是自定义的日志生成实现类的存放路径；mapper 是 HBase 中的表的 POJO 的实现类的存放路径，里面包含从数据库取出的 byte 型的数据向字符串和 Long 型的转换；service 是业务逻辑实现类包的存放路径；utils 是自定义工具包的存放路径，主要实现生成 RowKey 的实现方法；vo 是实体类的类包存放路径。用户消费子系统类包存放路径如图 13-4 所示。

（4）核心代码实现。

外部调用过程和商户子系统的类似，只是存取数据库采用的是 HBase。系统数据库访问层采用 spring.boot.starter.hbase 集成 HBase，spring.boot.starter.hbase 为 HBase 的查询和更新等操作提供了简易的 API。系统通过侦听 Kafka，得到商户投放优惠券消息，并向 HBase 数据库中的优惠券表（pb:passtemplate）中插入数据。用户消费子系统运行流程如图 13-5 所示。

图 13-4 用户消费子系统类包存放路径

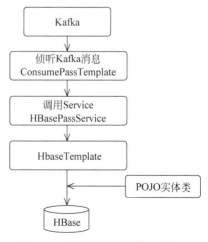

图 13-5 用户消费子系统运行流程

① Service 层的 serviceConsumePassTemplate 类，主要实现侦听 Kafka 消息及调用 HBasePassService 向 HBase 插入数据，代码如下：

```
package com.coupon.passbook.service;
```

```java
import com.alibaba.fastjson.JSON;
import com.coupon.passbook.vo.PassTemplate;
import com.coupon.passbook.constant.Constants;
import lombok.extern.slf4j.Slf4j;
import org.springframework.beans.factory.annotation.Autowired;
import org.springframework.kafka.annotation.KafkaListener;
import org.springframework.kafka.support.KafkaHeaders;
import org.springframework.messaging.handler.annotation.Header;
import org.springframework.messaging.handler.annotation.Payload;
import org.springframework.stereotype.Component;
/**
 * <h1>消费 Kafka 中的 PassTemplate</h1>
 * Created by hhao.
 */
@Slf4j
@Component
public class ConsumePassTemplate{
    /** pass 相关的 HBase 服务 */
    private final IHBasePassService passService;
    @Autowired
    public ConsumePassTemplate(IHBasePassService passService){
        this.passService = passService;
    }
    //监听 Kafka 消息,并将商户投放的优惠券存入 HBase 数据库
    @KafkaListener(topics = {Constants.TEMPLATE_TOPIC})
    public void receive(@Payload String passTemplate,
    //Kafka 消息 key 值
    @Header(KafkaHeaders.RECEIVED_MESSAGE_KEY) String key,
    //Kafka 消息 partition 值
    @Header(KafkaHeaders.RECEIVED_PARTITION_ID) int partition ,
    //Kafka 消息 topic 值
    @Header(KafkaHeaders.RECEIVED_TOPIC) String topic) {
        log.info("Consumer Receive PassTemplate: {}", passTemplate);
        PassTemplate pt;
        try{
           //格式转换
           pt = JSON.parseObject(passTemplate, PassTemplate.class);
        } catch (Exception ex) {
            log.error("Parse PassTemplate Error: {}",ex.getMessage());
            return;
        }
        //调用 HBasePassService 向 HBase 插入数据
        passService.dropPassTemplateToHBase(pt);
    }
}
```

② Server 层的 HBasePassService 类,主要功能是利用 HBase 客户端 HbaseTemplate 向 Hbase 插入优惠券数据,代码如下:

```java
package com.coupon.passbook.service.impl;
import com.coupon.passbook.constant.Constants;
import com.coupon.passbook.service.IHBasePassService;
```

```java
import com.coupon.passbook.utils.RowKeyGenUtil;
import com.coupon.passbook.vo.PassTemplate;
import com.spring4all.spring.boot.starter.hbase.api.HbaseTemplate;
import lombok.extern.slf4j.Slf4j;
import org.apache.commons.lang.time.DateFormatUtils;
import org.apache.hadoop.hbase.TableName;
import org.apache.hadoop.hbase.client.Get;
import org.apache.hadoop.hbase.client.Put;
import org.apache.hadoop.hbase.util.Bytes;
import org.springframework.beans.factory.annotation.Autowired;
import org.springframework.stereotype.Service;
/**
 * <h1> Pass HBase 服务</h1>
 *  .
 */
@Slf4j
@Service
//向 HBase 插入优惠券数据
public class HBasePassServiceImpl implements IHBasePassService{
    /** HBase 客户端 */
    @Autowired
    private final HbaseTemplate hbaseTemplate;
        public HBasePassServiceImpl(HbaseTemplate hbaseTemplate){
            this.hbaseTemplate = hbaseTemplate;
}
//生成优惠券行键
@Override
    public boolean dropPassTemplateToHBase(PassTemplate passTemplate){
        if (null == passTemplate){
            return false;
        }
        //生成优惠券行键
        String rowKey = RowKeyGenUtil.genPassTemplateRowKey(passTemplate);
        //连接 HBase 数据库,并判断行键是否存在
        try {
            if
            (hbaseTemplate.getConnection().getTable(TableName.valueOf( Constants.
            PassTemplateTable.TABLE_NAME)).exists(new Get(Bytes.toBytes(rowKey)))){
                log.warn("RowKey {} is already exist!",rowKey);
                return false;
            }
        } catch (Exception ex) {
            log.error("DropPassTemplateToHBase Error: {}",ex.getMessage());
            return false;
        }
        //插入行键
        Put put = new Put(Bytes.toBytes(rowKey));
        //插入优惠券 id
        put.addColumn(
```

```java
                Bytes.toBytes(Constants.PassTemplateTable.FAMILY_B),
                Bytes.toBytes(Constants.PassTemplateTable.ID),
                Bytes.toBytes(passTemplate.getId())
);
//插入优惠券名称
put.addColumn(
        Bytes.toBytes(Constants.PassTemplateTable.FAMILY_B),
        Bytes.toBytes(Constants.PassTemplateTable.TITLE),
        Bytes.toBytes(passTemplate.getTitle())
);
//插入优惠券摘要
put.addColumn(
        Bytes.toBytes(Constants.PassTemplateTable.FAMILY_B),
        Bytes.toBytes(Constants.PassTemplateTable.SUMMARY),
);
//插入优惠券详细信息
put.addColumn(
        Bytes.toBytes(Constants.PassTemplateTable.FAMILY_B),
        Bytes.toBytes(Constants.PassTemplateTable.DESC),
        Bytes.toBytes(passTemplate.getDesc())
);
//插入优惠券 Token
put.addColumn(
        Bytes.toBytes(Constants.PassTemplateTable.FAMILY_B),
        Bytes.toBytes(Constants.PassTemplateTable.HAS_TOKEN),
        Bytes.toBytes(passTemplate.getHasToken())
);
//插入优惠券背景色
put.addColumn(
        Bytes.toBytes(Constants.PassTemplateTable.FAMILY_B),
        Bytes.toBytes(Constants.PassTemplateTable.BACKGROUND),
        Bytes.toBytes(passTemplate.getBackground())
);
//插入优惠券数量
put.addColumn(
        Bytes.toBytes(Constants.PassTemplateTable.FAMILY_C),
        Bytes.toBytes(Constants.PassTemplateTable.LIMIT),
        Bytes.toBytes(passTemplate.getLimit())
);
//优惠券开始时间
put.addColumn(
        Bytes.toBytes(Constants.PassTemplateTable.FAMILY_C),
        Bytes.toBytes(Constants.PassTemplateTable.START),
        Bytes.toBytes(DateFormatUtils.ISO_DATE_FORMAT.format(passTemplate.
        getStart()))
);
//插入优惠结束时间
put.addColumn(
        Bytes.toBytes(Constants.PassTemplateTable.FAMILY_C),
```

```
            Bytes.toBytes(Constants.PassTemplateTable.END),
            Bytes.toBytes(DateFormatUtils.ISO_DATE_FORMAT.format(passTemplate.
            getEnd()))
    );
    //提交数据

    hbaseTemplate.saveOrUpdate(Constants.PassTemplateTable.TABLE_NAME,put);
        return true;
    }
}
```

13.5 系统运行测试

13.5.1 启动系统

（1）启动 HBase 数据库，并在 HBase 数据库下新建表，新建表语句如下：

```
create_namespace 'pb'
create 'pb:user','{NAME => 'b', VERSIONS => '3',TTL => '2147483647','BLOOMFILTER' =>'ROW'},
'{NAME => 'o', VERSIONS => '3', TTL => '2147483647',' BLOOMFILTER' => 'ROW'}
create 'pb:pass','{NAME => 'i', VERSIONS => '3', TTL => '2147483647', BLOOMFILTER' => 'ROW'}
create 'pb:passtemplate','{NAME => 'b', VERSIONS => '3', TTL =>'2147483647','BLOOMFILTER' => 'ROW'}, {NAME => 'c', VERSIONS => '3', TTL => '2147483647','BLOOMFILTER' => 'ROW'}
```

（2）启动 MySQL 数据库，并在 MySQL 数据库下新建商户表，新建表语句如下：

```
CREATE TABLE `merchants`(
  `id` int(10)unsigned NOT NULL AUTO_INCREMENT,
  `name` varchar(64) COLLATE utf8_bin NOT NULL COMMENT '商户名称',
  `logo_url` varchar(256) COLLATE utf8_bin NOT NULL COMMENT '商户 logo',
  `business_license_url` varchar(256) COLLATE utf8_bin NOT NULL COMMENT '营业执照',
  `phone` varchar(64) COLLATE utf8_bin NOT NULL COMMENT '商户联系电话',
  `address` varchar(64) COLLATE utf8_bin NOT NULL COMMENT '商户地址',
  `is_audit` BOOLEAN NOT NULL COMMENT '是否通过审核',
  PRIMARY KEY (`id`)
) ENGINE = InnoDB AUTO_INCREMENT = 17 DEFAULT CHARSET = utf8;
```

（3）安装 Redis 并启动。

（4）安装 Kafka 并启动。

（5）启动商户投放子系统，检查 application.yml 中的 MySQL 数据库和 Kafka 的连接是否配置正确。其中，Kafka 中的 bootstrap-servers 最好配置成主机名，并和 Kafka 服务端的 server.properties 中配置的主机名相同，同时将客户端和服务端的主机名在 hosts 文件中配置好 IP 和主机名的映射关系。找到程序的主启动类 MerchantsApplication，右击后，在弹出的快捷菜单中选择"启动"选项。商户投放子系统启动成功信息如图 13-6 所示。

（6）启动用户消费子系统，检查 application.yml 中的 MySQL、HBase 数据库和 Kafka 的连接是否配置正确。其中，Kafka 配置中的 bootstrap-servers 最好配置成主机

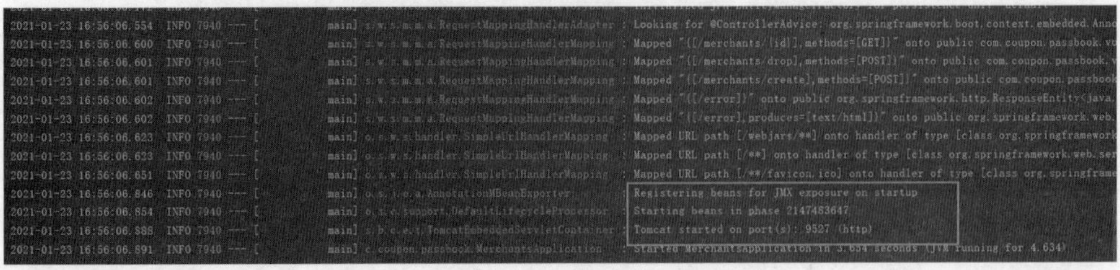

图 13-6　商户投放子系统启动成功信息

名,并和 Kafka 服务端的 server.properties 中配置的主机名相同,同时将客户端和服务端的主机名在 hosts 文件中配置好 IP 和主机名的映射关系。HBase 配置中,quorum 配置成 Zookeeper 的主机名和默认端口号。HBase 安装时最好不要选择启动 HBase 内部集成的 Zookeeper,应安装一个外部的 Zookeeper 服务,以便 HBase 和 Kafka 可以共用相同的 Zookeeper 服务。找到程序的主启动类 PassbookApplication,右击后,在弹出的快捷菜单中选择"启动"选项。用户消费子系统启动成功信息如图 13-7 所示。

图 13-7　用户消费子系统启动成功信息

13.5.2　商户投放子系统测试

使用 IntelliJ IDEA tools 下的 HTTP client 的 test RESTful Web Service 工具作为客户端调用工具进行系统测试,主要过程如下。

(1) 创建商户,调用 merchants/create()方法创建商户基本信息。调用参数如下:

```
POST http://125.223.37.209:9527/merchants/create
Accept: */*
token: passbook-merchants
Content-Type: application/json;charset=UTF-8
{
  "name": "大数据教程",
  "logoUrl": "www.bigdata.com",
  "businessLicenseUrl": "www.bigdata.com",
  "phone": "1234567890",
  "address": "哈尔滨"
}
```

调用成功后,创建商户并返回结果界面,如图 13-8 所示。返回码为 errorCode=0,表示创建成功,创建 id 为 2 的一个商户。

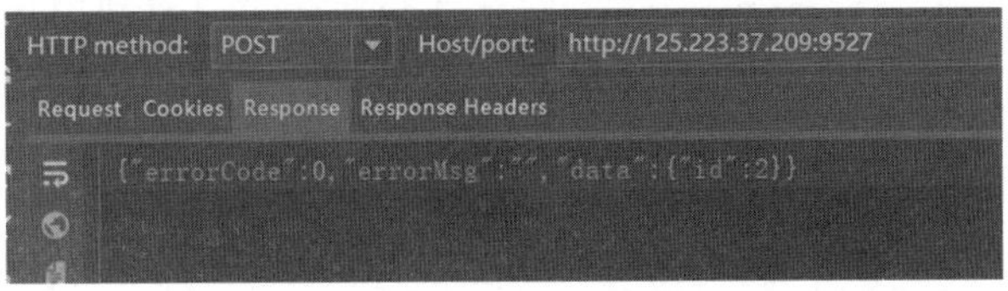

图 13-8 创建商户并返回结果界面

（2）发放优惠券。调用 merchants/drop( )方法发放优惠券，JSON 中的数据项 id 要与创建的商户 id 一致。调用参数如下：

```
POST http://125.223.37.209:9527/merchants/drop
Accept: */*
token: passbook-merchants
Content-Type: application/json;charset=UTF-8
{
    "background": 1,
    "desc": "大数据优惠券",
    "end": "2021-06-30",
    "hasToken": false,
    "id": 2,
    "limit": 1000,
    "start": "2021-01-01",
    "summary": "优惠券简介",
    "title": "大数据优惠券-1"
}
```

调用成功后，发放优惠券并返回结果界面，如图 13-9 所示。errorCode＝0 表示调用成功。

图 13-9 发放优惠券返回结果界面

商户投放子系统控制台日志信息如图 13-10 所示。可以看到已经将数据发送给 Kafka。

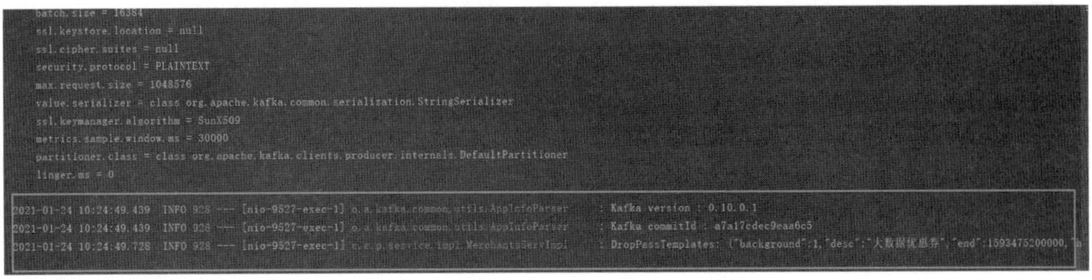

图 13-10 商户投放子系统控制台日志信息

用户消费子系统控制台日志信息如图 13-11 所示。可以看到用户消费子系统已经侦听到 Kafka 消息,并将接收到的消息存入 HBase 数据库中,生成的 Row Key 为 d159f46f-7112d7de25d92b24d4b392a0。

图 13-11　用户消费子系统控制台日志信息

利用 HBase Shell 查看 pb:passtemplate 表的数据,pb:passtemplate 表数据的插入情况如图 13-12 所示。可以看到 pb:passtemplate 表中已经被插入了数据。

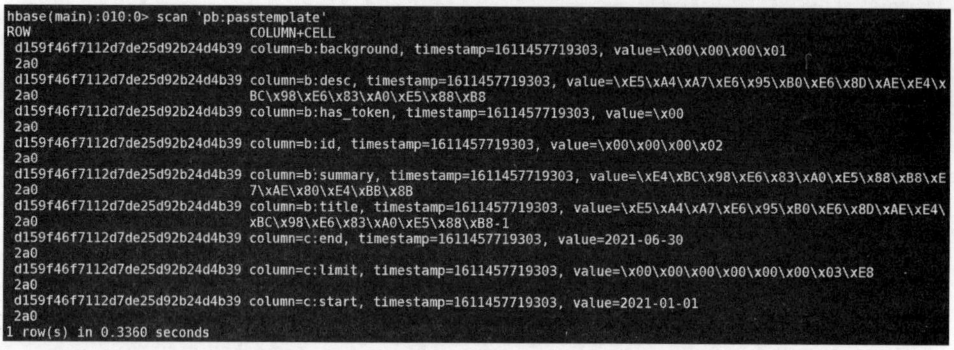

图 13-12　pb:passtemplate 表数据的插入情况

13.5.3　用户消费子系统测试

(1) 创建消费用户,调用 passbook/createuser()方法创建用户基本信息。调用参数如下:

```
POST http://125.223.37.209:9528/passbook/createuser
Accept: */*
Content-Type: application/json;charset=UTF-8
{
    "baseInfo":{
        "name": "hhao",
        "age": 20,
        "sex": "m"
    },
    "otherInfo":{
        "phone": "13313678445",
        "address": "harbin"
    }
}
```

调用成功后,创建消费用户并返回结果界面,如图 13-13 所示。errorCode＝0 表示调用成功,创建了一个 id 为 105065 的消费用户。

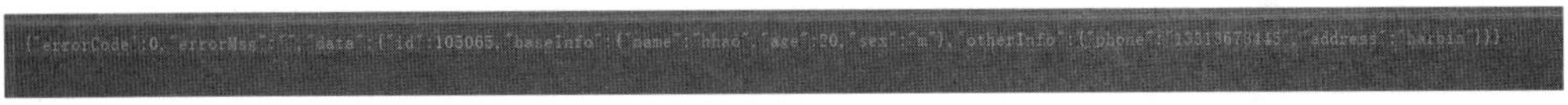

```
{"errorCode":0,"errorMsg":"","data":{"id":105065,"baseInfo":{"name":"hhao","age":20,"sex":"m"},"otherInfo":{"phone":"13313678445","address":"harbin"}}}
```

图 13-13　创建消费用户并返回结果界面

利用 HBase Shell 查看 pb:user 表的数据，pb:user 表数据的插入情况如图 13-14 所示。可以看到 pb:user 表中已经插入了数据。

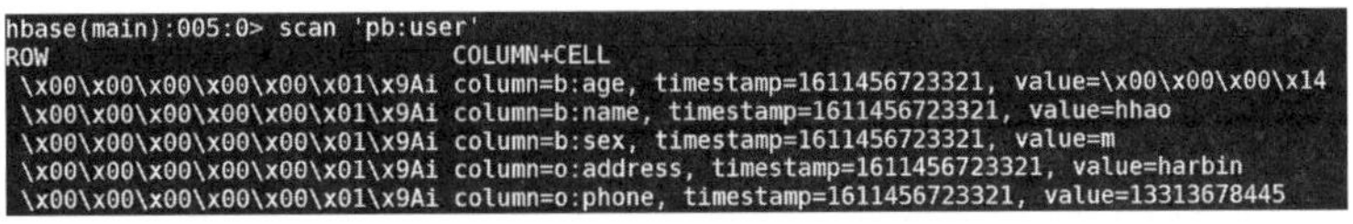

```
hbase(main):005:0> scan 'pb:user'
ROW                              COLUMN+CELL
 \x00\x00\x00\x00\x00\x01\x9Ai column=b:age, timestamp=1611456723321, value=\x00\x00\x00\x14
 \x00\x00\x00\x00\x00\x01\x9Ai column=b:name, timestamp=1611456723321, value=hhao
 \x00\x00\x00\x00\x00\x01\x9Ai column=b:sex, timestamp=1611456723321, value=m
 \x00\x00\x00\x00\x00\x01\x9Ai column=o:address, timestamp=1611456723321, value=harbin
 \x00\x00\x00\x00\x00\x01\x9Ai column=o:phone, timestamp=1611456723321, value=13313678445
```

图 13-14　pb:user 表数据的插入情况

(2) 领取优惠券，调用 passbook/gainpasstemplate()方法创建用户基本信息。调用参数如下：

```
POST http://125.223.37.209:9528/passbook/gainpasstemplate
Accept: */*
Content-Type: application/json;charset=UTF-8
 {
     "userId": 105065,
     "passTemplate":{
         "id": 2,
         "title": "大数据优惠券-1",
         "hasToken": false
     }
 }
```

errorCode=0 表示调用成功，表示消费用户 id 为 105065 的用户已经领取了优惠券。

利用 HBase Shell 查看 pb:pass 表的数据，pb:pass 表的数据插入情况如图 13-15 所示。可以看到 pb:pass 表中已经插入了数据。

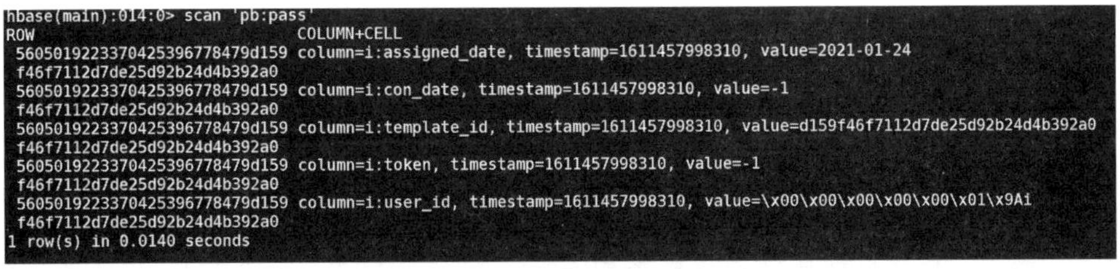

```
hbase(main):014:0> scan 'pb:pass'
ROW                              COLUMN+CELL
 56050192233704253967784 79d159 column=i:assigned_date, timestamp=1611457998310, value=2021-01-24
 f46f7112d7de25d92b24d4b392a0
 560501922337042539677847 9d159 column=i:con_date, timestamp=1611457998310, value=-1
 f46f7112d7de25d92b24d4b392a0
 560501922337042539677847 9d159 column=i:template_id, timestamp=1611457998310, value=d159f46f7112d7de25d92b24d4b392a0
 f46f7112d7de25d92b24d4b392a0
 560501922337042539677847 9d159 column=i:token, timestamp=1611457998310, value=-1
 f46f7112d7de25d92b24d4b392a0
 560501922337042539677847 9d159 column=i:user_id, timestamp=1611457998310, value=\x00\x00\x00\x00\x00\x01\x9Ai
 f46f7112d7de25d92b24d4b392a0
1 row(s) in 0.0140 seconds
```

图 13-15　pb:pass 表的数据插入情况

13.6　本章小结

本章介绍了分布式优惠券后台应用系统的开发核心思路和核心代码以及测试调用过程，系统分为商户优惠券发放子系统和用户消费子系统。由于优惠券系统比较庞大，代码较多，本书只列出了系统的基本构建和核心思路与代码。全部代码请查看本书的配套资源。

第 14 章

实战案例——新闻话题实时统计分析系统

14.1 系统简介

新闻话题实时统计分析系统以搜狗实验室的用户查询日志为基础，模拟生成用户查询日志，通过 Flume 将日志进行实时采集、汇集、分析并进行存储。利用 Spark Streaming 实时统计分析前 20 名流量最高的新闻话题，并在前端页面实时显示结果。

14.2 系统总体架构

系统总体架构如图 14-1 所示。

（1）利用搜狗实验室的用户查询日志通过模拟日志生成程序生成用户查询日志，供 Flume 采集。

（2）日志采集端 Flume 采集数据发送给 Flume 日志汇聚节点，并进行预处理。

（3）Flume 将预处理的数据进行数据存储，存储到 HBase 数据库中，并发送消息给 Kafka 的 Topic。

（4）Spark Streaming 接收 Kafka 的 Topic 实时消息并计算实时话题的数量，并将计算结果保存到 MySQL 数据库中。

（5）前端页面通过建立 WebSocket 通道读取 MySQL 数据库中的数据，实时展示话题的动态变化。

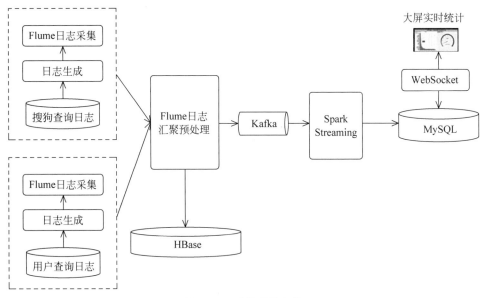

图 14-1　系统总体架构

14.3　表结构设计

（1）MySQL 的表结构设计。webCount（新闻话题数量表）如表 14-1 所示。

表 14-1　webCount（新闻话题数量表）

字　段　名	类　　型	大　　小	是否主键	说　　明
id	int	10	是	话题 ID
titleName	varchar	255		话题名称
count	int	11		话题数量

（2）HBase 的表结构设计。weblogs（日志表）如表 14-2 所示。

表 14-2　weblogs（日志表）

基本信息列族（info）	
datetime	查询时间
userid	用户 ID
searchname	搜索词
retorder	URL 在返回结果中的排名
cliorder	用户点击的顺序号
cliurl	用户点击的 URL

14.4 系统实现

14.4.1 模拟日志生成程序

（1）在 IntelliJ IDEA 构建 Java 项目 weblogs。编写数据生成模拟类，其主要功能是读取搜狗用户日志文件，并构建新的格式写入一个新文件供 Flume 采集，代码如下：

```java
package main.java;
import java.io.*;
public class ReadWrite {
    static String readFileName;
    static String writeFileName;
    //主函数
    public static void main(String args[]){
        //读入文件路径,为搜狗用户日志文件路径
        readFileName = args[0];
        //写入文件路径
        writeFileName = args[1];
        try {
            //调用自定义方法按行读取数据,并写入模拟日志文件
            readFileByLines(readFileName);
        }catch(Exception e){
        }
    }
    //按行读取数据,并写入模拟日志文件,输入参数为文件路径
    public static void readFileByLines(String fileName) {
        FileInputStream fis = null;
        InputStreamReader isr = null;
        BufferedReader br = null;
        String tempString = null;
        try {
            System.out.println("以行为单位读取文件内容,一次读一整行: ");
            fis = new FileInputStream(fileName);
            //以文件流和GBK编码格式打开文件
            isr = new InputStreamReader(fis,"GBK");
            //获取文件中的内容
            br = new BufferedReader(isr);
            int count = 0;
            //按行读取
            while
            ((tempString = br.readLine()) != null) {
                    //统计行数
                    count++;
                //延时300ms
                Thread.sleep(300);
                //字符集转换
                String str = new String(tempString.getBytes("UTF8"),"GBK");
                System.out.println("row:" + count + ">>>>>>>>" + tempString);
```

```
                //调用自定义写文件方法
                appendMethodA(writeFileName,tempString);
            }
            isr.close();
        }
        catch (IOException e) {
            e.printStackTrace();
        } catch (InterruptedException e) {
            e.printStackTrace();
        } finally {
            if (isr!=null) {
                try {
                    isr.close();
                } catch (IOException e1) {
                }
            }
        }
    }
    //自定义写文件方法,输入参数为文件名、写入内容
    public static void appendMethodA (String file,String conent) {
        BufferedWriter out = null;
        try {
            //生成写入文件,没有文件则建立文件,有文件则打开文件
            out = new BufferedWriter(new OutputStreamWriter(
                    new FileOutputStream(file, true)));
            //向文件写入回车符
            out.write("\n");
            //向文件写入内容
            out.write(conent);
        }
        catch (Exception e) {
            e.printStackTrace();
        } finally {
            try {
                //关闭文件
                out.close();
            }
            catch (IOException e) {
                e.printStackTrace();
            }
        }
    }
}
```

(2) 生成 JAR 包,并将 JAR 包上传到生成日志服务器。生成 JAR 包界面如图 14-2 所示。

(3) 编写 weblog.sh,调用模拟日志生成 JAR 包,并将 weblog.sh 上传到生成日志服务器。代码如下:

```
#/bin/bash
echo "start log......"
java -jar /opt/jars/weblog.jar /opt/datas/weblog.log /opt/datas/weblog-flume.log
```

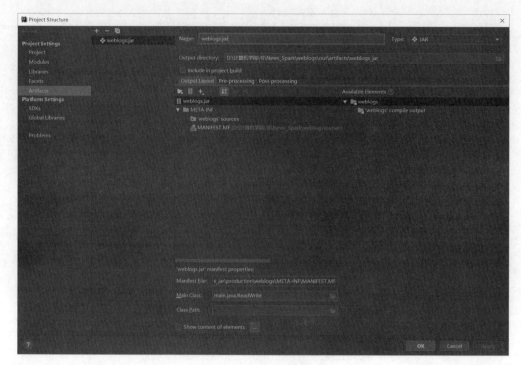

图 14-2　生成 JAR 包界面

14.4.2　Flume 配置

(1) 配置日志采集端的 Flume 服务,新建 flume-weblog-conf. properties 文件,配置如下:

```
a2.sources = r1
a2.sinks = k1
a2.channels = c1
a2.sources.r1.type = exec
a2.sources.r1.command = tail - F /opt/datas/weblog - flume.log
a2.sources.r1.channels = c1
a2.channels.c1.type = memory
a2.channels.c1.capacity = 1000
a2.channels.c1.transactionCapacity = 1000
a2.channels.c1.keep - alive = 5
a2.sinks.k1.type = avro
a2.sinks.k1.channel = c1
a2.sinks.k1.hostname = dc3 - data
a2.sinks.k1.port = 5555
```

其中,a2.sources.r1.command 中配置模拟生成日志的路径,a2.sinks.k1.hostname 是日志汇聚端服务器的主机名。注意,在 hosts 文件中要配置主机名和 IP 的映射关系。

(2) 配置日志汇聚端的 Flume 服务,新建 flume-weblog-conf.properties,配置如下：

```
a1.sources = r1
a1.channels = kafkaC hbaseC
a1.sinks = kafkaSink hbaseSink
a1.sources.r1.type = avro
a1.sources.r1.channels = hbaseC kafkaC
a1.sources.r1.bind = dc3-data
a1.sources.r1.port = 5555
a1.sources.r1.threads = 5
# *************************** Flume + HBase ******************************
a1.channels.hbaseC.type = memory
a1.channels.hbaseC.capacity = 10000
a1.channels.hbaseC.transactionCapacity = 10000
a1.channels.hbaseC.keep-alive = 20
a1.sinks.hbaseSink.type = asynchbase
a1.sinks.hbaseSink.table = weblogs
a1.sinks.hbaseSink.columnFamily = info
a1.sinks.hbaseSink.serializer = org.apache.flume.sink.hbase.KfkAsyncHbaseEventSerializer
a1.sinks.hbaseSink.channel = hbaseC
a1.sinks.hbaseSink.serializer.payloadColumn = datetime,userid,searchname,retorder,cliorder,cliurl
# *************************** Flume + Kafka ******************************
a1.channels.kafkaC.type = memory
a1.channels.kafkaC.capacity = 10000
a1.channels.kafkaC.transactionCapacity = 10000
a1.channels.kafkaC.keep-alive = 20
a1.sinks.kafkaSink.channel = kafkaC
a1.sinks.kafkaSink.type = org.apache.flume.sink.kafka.KafkaSink
a1.sinks.kafkaSink.brokerList = dc3-data:9092
a1.sinks.kafkaSink.topic = weblogs
a1.sinks.kafkaSink.zookeeperConnect = dc3-data:2181
a1.sinks.kafkaSink.requiredAcks = 1
a1.sinks.kafkaSink.batchSize = 1
a1.sinks.kafkaSink.serializer.class = kafka.serializer.StringEncoder
```

其中,a1.sinks.hbaseSink.table 是 HBase 数据库的表名,a1.sinks.hbaseSink.columnFamily 是 HBase 数据库的表的列族名,a1.sinks.hbaseSink.serializer.payloadColumn 是列族中的字段名,a1.sources.r1.bind 是日志汇聚端服务器的主机名,a1.sinks.kafkaSink.brokerList 是配置 Kafka 服务器的主机名和端口号。注意,在 hosts 文件中要配置主机名和 IP 的映射关系。

(3) 自定义 SinkHBase 程序设计与开发。

① 将 apache-flume-1.7.0-src.tar.gz 源码下载到本地解压。

② 导入 flume-ng-hbase-sink 源码,打开 IntelliJ IDEA,选择 File→Open 选项,选中 flume-ng-hbase-sink,单击 OK 按钮加载相应模块的源码。导入 flume-ng-hbase-sink 源码界面如图 14-3 所示。

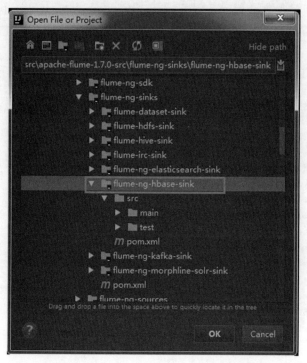

图 14-3 导入 flume-ng-hbase-sink 源码界面

（3）模仿 SimpleAsyncHbaseEventSerializer 自定义 KfkAsyncHbaseEventSerializer 实现类，只修改 getActions()方法，代码如下：

```
@Override
public List<PutRequest> getActions() {
    List<PutRequest> actions = new ArrayList<PutRequest>();
    if (payloadColumn != null) {
        byte[] rowKey;
        try {
            /* ---------------------- 代码修改开始 ---------------------- */
            //解析列字段
            String[] columns = new String(this.payloadColumn).split(",");
            //解析 Flume 采集过来的每行的值
            String[] values = new String(this.payload).split(",");
            for(int i = 0;i < columns.length;i++){
                byte[] colColumn = columns[i].getBytes();
                byte[] colValue = values[i].getBytes(Charsets.UTF_8);
                //数据校验：字段和值是否对应
                if (colColumn.length != colValue.length) {
                    break;
                }
                //时间
                String datetime = values[0].toString();
                //用户 id
                String userid = values[1].toString();
```

```
            //根据业务自定义 RowKey
            rowKey = SimpleRowKeyGenerator.getKfkRowKey(userid,datetime);
            //插入数据
            PutRequest putRequest = new PutRequest(table, rowKey,cf, colColumn,
            colValue);
            actions.add(putRequest);
            /* ---------------------- 代码修改结束 ---------------------- */
          }
        }
      catch (Exception e) {
        throw new FlumeException("Could not get row key!" , e);
      }
    }
    return actions;
}
```

(4) 修改 SimpleRowKeyGenerator 类，根据具体业务自定义 RowKey 生成方法，代码如下：

```
public static byte[] getKfkRowKey(String userid, String datetime) throws
  UnsupportedEncodingException {
    return (userid + "-" + datetime + "-" +
    String.valueOf(System.currentTimeMillis())).getBytes("UTF8");
}
```

(5) 生成 JAR 包，删除其他依赖包，只把 flume-ng-hbase-sink 打成 JAR 包（见框内）。生成 JAR 包界面如图 14-4 所示。

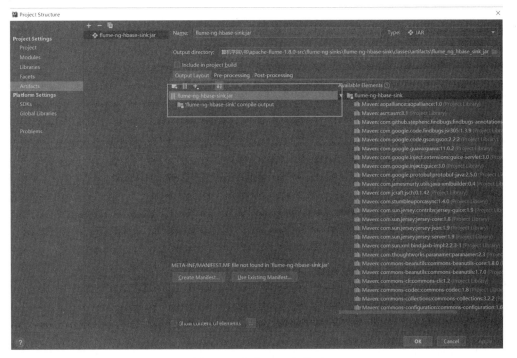

图 14-4　生成 JAR 包界面

(6) 上传 JAR 包。将打包的名字替换为 Flume 默认的包名 flume-ng-hbase-sink-1. 7.0.jar,然后上传至日志汇聚服务器上的 flume/lib 目录下,覆盖原有的 JAR 包。

14.4.3 配置 Kafka

配置 server.properties 文件,其中 listeners 是配置 Kafka 服务器的主机名和端口号, zookeeper.connect 是配置 Zookeeper 的主机名和端口号。注意,在 hosts 文件中要配置主机名和 IP 的映射关系。配置如下:

```
listeners = PLAINTEXT://dc3-data:9092
zookeeper.connect = localhost:2181
```

14.4.4 Spark Streaming 开发

(1) 新建一个 MAVEN 工程,添加依赖包。pom.xml 文件内容如下:

```
<project xmlns = http://maven.apache.org/POM/4.0.0 xmlns:xsi = "http://www.w3.org/2001/XMLSchema-instance"
xsi:schemaLocation = "http://maven.apache.org/POM/4.0.0 http://maven.apache.org/maven-v4_0_0.xsd">
<modelVersion>4.0.0</modelVersion>
<groupId>com.spark</groupId>
<artifactId>sparkStu</artifactId>
<packaging>war</packaging>
<version>1.0-SNAPSHOT</version>
<name>sparkStu Maven Webapp</name>
<url>http://maven.apache.org</url>
<properties>
<hadoop.version>2.5.0</hadoop.version>
<scala.binary.version>2.11</scala.binary.version>
<spark.version>2.2.0</spark.version>
</properties>
<dependencies>
<dependency>
<groupId>org.apache.spark</groupId>
<artifactId>spark-core_${scala.binary.version}</artifactId>
<version>${spark.version}</version>
</dependency>
<dependency>
<groupId>org.apache.spark</groupId>
<artifactId>spark-sql_${scala.binary.version}</artifactId>
<version>${spark.version}</version>
</dependency>
<dependency>
<groupId>org.apache.spark</groupId>
<artifactId>spark-streaming_${scala.binary.version}</artifactId>
<version>${spark.version}</version>
</dependency>
<dependency>
```

```xml
<groupId>org.apache.spark</groupId>
<artifactId>spark-hive_${scala.binary.version}</artifactId>
<version>${spark.version}</version>
</dependency>
<dependency>
<groupId>org.apache.spark</groupId>
<artifactId>spark-streaming-kafka-0-8_${scala.binary.version}</artifactId>
<version>${spark.version}</version>
</dependency>
<dependency>
<groupId>org.apache.spark</groupId>
<artifactId>spark-streaming-kafka-0-10_${scala.binary.version}</artifactId>
<version>${spark.version}</version>
</dependency>
<dependency>
<groupId>org.apache.spark</groupId>
<artifactId>spark-sql-kafka-0-10_${scala.binary.version}</artifactId>
<version>${spark.version}</version>
</dependency>
<dependency>
<groupId>mysql</groupId>
<artifactId>mysql-connector-java</artifactId>
<version>5.1.35</version>
</dependency>
<dependency>
<groupId>org.apache.hadoop</groupId>
<artifactId>hadoop-client</artifactId>
<version>${hadoop.version}</version>
</dependency>
<dependency>
<groupId>com.alibaba</groupId>
<artifactId>fastjson</artifactId>
<version>1.2.33</version>
</dependency>
</dependencies>
<build>
<finalName>sparkStu</finalName>
</build>
</project>
```

（2）编写 Scala 类 StructuredStreamingKafka，实现从 Kafka 中读取数据存储到关系数据库 MySQL。代码如下：

```
package com.spark.test
import org.apache.spark.sql.SparkSession
import org.apache.spark.sql.streaming.ProcessingTime
/**
  * 结构化流从 Katka 中读取数据存储到关系数据库 MySQL
  * 目前结构化流对 Kafka 的要求版本 0.10 及以上
  */
```

```scala
object StructuredStreamingKafka{
  case class Weblog(datatime:String,
                    userid:String,
                    searchname:String,
                    retorder:String,
                    cliorder:String,
                    cliurl:String)
  //主函数
  def main(args: Array[String]): Unit = {
    val spark   = SparkSession.builder()
     .master("local[2]")
     .appName("streaming").getOrCreate()
      //从 Kafka 获取数据
    val df = spark
     .readStream
     .format("kafka")
     .option("kafka.bootstrap.servers","dc3-data:9092")
     .option("subscribe", "weblogs")
     .load()
   import spark.implicits._
    //数据处理
    val lines = df.selectExpr("CAST(value AS STRING)").as[String]
    //以逗号分割数据
    val weblog = lines.Map(_.split(", "))
                  .Map(x => Weblog(x(0),x(1),x(2),x(3),x(4),x(5)))
    //以查询词进行分组,计算数量
    val titleCount = weblog
     .groupBy("searchname").count().toDF("titleName","count")
//连接 MySQL 数据库 URL
val url = "jdbc:mysql://localhost:3306/test"
//MySQL 用户名
val username = "root"
//MySQL 密码
val password = "123456"
//连接 MySQL
val writer = new JDBCSink(url,username,password)
  //数据保存到 MySQL
  val query = titleCount.writeStream
    .foreach(writer)
    .outputMode("update")
    .trigger(ProcessingTime("5 seconds"))
    .start()
   query.awaitTermination()
  }
}
```

14.4.5 WebSocket 和前端界面开发

(1) 新建 pom.xml 文件,内容如下:

```xml
<?xml version = "1.0" encoding = "UTF-8"?>
```

```xml
<project xmlns="http://maven.apache.org/POM/4.0.0"
  xmlns:xsi="http://www.w3.org/2001/XMLSchema-instance"
  xsi:schemaLocation="http://maven.apache.org/POM/4.0.0
  http://maven.apache.org/xsd/maven-4.0.0.xsd">
  <modelVersion>4.0.0</modelVersion>
  <groupId>com.spark.service</groupId>
  <artifactId>spark_socket</artifactId>
  <version>1.0-SNAPSHOT</version>
  <packaging>war</packaging>
  <name>spark_socket Maven Webapp</name>
  <!-- FIXME change it to the project's website -->
  <url>http://www.example.com</url>
  <properties>
    <project.build.sourceEncoding>UTF-8</project.build.sourceEncoding>
    <maven.compiler.source>1.7</maven.compiler.source>
    <maven.compiler.target>1.7</maven.compiler.target>
  </properties>
  <dependencies>
    <dependency>
      <groupId>junit</groupId>
      <artifactId>junit</artifactId>
      <version>4.11</version>
      <scope>test</scope>
    </dependency>
    <dependency>
      <groupId>javax</groupId>
      <artifactId>javaee-api</artifactId>
      <version>8.0</version>
      <scope>provided</scope>
    </dependency>
    <dependency>
      <groupId>junit</groupId>
      <artifactId>junit</artifactId>
      <version>3.8.1</version>
      <scope>test</scope>
    </dependency>
    <dependency>
      <groupId>com.alibaba</groupId>
      <artifactId>fastjson</artifactId>
      <version>1.2.33</version>
    </dependency>
  </dependencies>
  <build>
    <finalName>spark_socket</finalName>
    <pluginManagement>
      <plugins>
        <plugin>
          <artifactId>maven-clean-plugin</artifactId>
          <version>3.1.0</version>
```

```xml
        </plugin>
        <plugin>
          <artifactId>maven-resources-plugin</artifactId>
          <vesion>3.0.2</version>
        </plugin>
        <plugin>
          <artifactId>maven-compiler-plugin</artifactId>
          <version>3.8.0</version>
        </plugin>
        <plugin>
          <artifactId>maven-surefire-plugin</artifactId>
          <version>2.22.1</version>
        </plugin>
        <plugin>
          <artifactId>maven-war-plugin</artifactId>
          <version>3.2.2</version>
        </plugin>
        <plugin>
          <artifactId>maven-install-plugin</artifactId>
          <version>2.5.2</version>
        </plugin>
        <plugin>
          <artifactId>maven-deploy-plugin</artifactId>
          <version>2.8.2</version>
        </plugin>
      </plugins>
    </pluginManagement>
  </build>
</project>
```

(2) 编写 Java 类 WeblogService,实现功能为连接 MySQL 数据库,取统计数据。代码如下:

```java
package com.spark.service;
import java.sql.Connection;
import java.sql.DriverManager;
import java.sql.PreparedStatement;
import java.sql.ResultSet;
import java.util.HashMap;
import java.util.Map;
/**
 * 连接 MySQL,取统计数据
 */
public class WeblogService{
    //MySQL 链接地址
    static String url
        = "jdbc:mysql://localhost:3306/test?useUnicode=true&characterEncoding=utf-8";
    //MySQL 用户名
    static String username = "root";
    //MySQL 密码
```

```java
static String password = "123456";
/**
 * 按话题分组,取数据
 */
public Map<String,Object> queryWeblogs(){
    Connection conn = null;
    PreparedStatement pst = null;
    String[] titleNames = new String[20];
    String[] titleCounts = new String[20];
    Map<String,Object> retMap = new HashMap<String,Object>();
    try{
        Class.forName("com.mysql.jdbc.Driver");
        //按配置连接 MySQL
        conn = DriverManager.getConnection(url,username,password);
        String query_sql = "select titleName,count from webCount where 1 = 1 order by count desc limit 20";
        //查询排名前 20 数据
        pst = conn.prepareStatement(query_sql);
        ResultSet rs = pst.executeQuery();
        int i = 0;
        while(rs.next()){
            //取得话题名称
            String titleName = rs.getString("titleName");
            //该话题名称的查询数据
            String titleCount = rs.getString("count");
            titleNames[i] = titleName;
            titleCounts[i] = titleCount;
            ++i;
        }
        retMap.put("titleName", titleNames);
        retMap.put("titleCount", titleCounts);
    }catch(Exception e){
        e.printStackTrace();
    }
    return retMap;
}
/**
 * 取话题总数
 */
public String[] titleCount(){
    Connection conn = null;
    PreparedStatement pst = null;
    String[] titleSums = new String[1];
    try{
        Class.forName("com.mysql.jdbc.Driver");
        //按配置连接 MySQL
        conn = DriverManager.getConnection(url,username,password);
        String query_sql = "select count(1) titleSum from webCount";
        //建立查询
```

```
                pst = conn.prepareStatement(query_sql);
                ResultSet rs = pst.executeQuery();
                if(rs.next()){
                    //话题数量
                    String titleSum = rs.getString("titleSum");
                    titleSums[0] = titleSum;
                }
            }catch(Exception e){
                e.printStackTrace();
            }
            return titleSums;
        }
    }
```

(3) 编写 Java 类 WeblogService，实现功能建立 WebSocket 通信，取统计数据供前端调用。代码如下：

```
package com.spark.service;
import com.alibaba.fastjson.JSON;
import javax.websocket.OnClose;
import javax.websocket.OnMessage;
import javax.websocket.OnOpen;
import javax.websocket.Session;
import javax.websocket.server.ServerEndpoint;
import java.io.IOException;
import java.util.HashMap;
import java.util.Map;
/**
 * 建立 WebSocket,供前端调用,取统计数据
 */
@ServerEndpoint("/websocket1")
public class WeblogSocket{
    WeblogService  weblogService = new WeblogService();
    @OnMessage
    public void onMessage(String message,Session session)
            throws IOException, InterruptedException {
        while(true){
            Map<String,Object> Map = new HashMap<String,Object>();
            //取得话题名称
            Map.put("titleName",weblogService.queryWeblogs().get("titleName"));
            //该话题名称的数量
            Map.put("titleCount",weblogService.queryWeblogs().get("titleCount"));
            //话题总数
            Map.put("titleSum",weblogService.titleCount());
            //发送 JSON 数据
            session.getBasicRemote().sendText(JSON.toJSONString(Map));
            //每 5s 去数据库获得数据
            Thread.sleep(5000);
            Map.clear();
        }
```

```
    }
    @OnOpen
    public void onOpen(){
        System.out.println("Client connected");
    }
        @OnClose
        public void onClose(){
            System.out.println("Connection closed");
        }
    }
```

(4) 建立大屏显示页面 index.html,实时进行大屏显示。在 web-inf 下新建目录 js,将 echarts.min.js 和 jquery-3.2.1.js 复制到该目录下。实现代码如下:

```
<!DOCTYPE html>
<html lang="en">
<head>
    <meta charset="UTF-8">
    <title>Title</title>
    <script src="js/echarts.min.js"></script>
    <script src="js/jquery-3.2.1.js"></script>
    <style>
        body{
            text-align:center;
            background-color:#dbdddd;
        }
        .div{ margin:0 auto; width:1000px; height:800px; border:1px solid #F00}
        /* CSS 注释:为了观察效果设置宽度、边框、高度等样式 */
    </style>
</head>
<body>
<h1>新闻话题用户浏览实时统计分析</h1>
<div>
    <div id="main" style="width:880px;height: 700px;float:left;">第一个</div>
    <div id="sum" style="width:800px;height: 700px;float:left;">第二个</div>
</div>

<div>
    <input type="submit" value="实时分析" onclick="start()" />
</div>
<div id="messages"></div>
<script type="text/javascript">
//WebSocket 访问地址,特别注意的是端口号和应用程序路径(此处为
//spark_socket_war_exploded)要根据应用服务的实际情况进行修改
    var webSocket = new
    WebSocket('ws://localhost:8080/spark_socket_war_exploded/websocket1');
    var myChart = echarts.init(document.getElementById('main'));
    var myChart_sum = echarts.init(document.getElementById('sum'));
    webSocket.onerror = function(event){
        onError(event)
```

```javascript
};
webSocket.onopen = function(event){
    onOpen(event)
};
webSocket.onmessage = function(event){
    onMessage(event)
};
function onMessage(event){
    var sd = JSON.parse(event.data);
    processingData(sd);
    titleSum(sd.titleSum);
}
function onOpen(event){
}
function onError(event){
    alert(event.data);
}
function start(){
    webSocket.send('hello');
    return false;
}
function processingData(json){
    var option = {
        backgroundColor: '#ffffff',          //背景色
        title:{
            text: '新闻话题浏览量【实时】排行',
            subtext: '数据来自搜狗实验室',
            textStyle:{
                fontWeight: 'normal',        //标题颜色
                color: '#408829'
            },
        },
        tooltip: {
            trigger:'axis',
            axisPointer:{
                type: 'shadow'
            }
        },
        legend: {
            data: ['浏览量']
        },
        grid: {
            left: '3%',
            right: '4%',
            bottom: '3%',
            containLabel: true
        },
        xAxis: {
            type: 'value',
```

```
                    boundaryGap: [0, 0.01]
                },
                yAxis: {
                    type: 'category',
                    data:json.titleName
                },
                series: [
                    {
                        name: '浏览量',
                        type: 'bar',
                        label:{
                            normal:{
                                show: true,
                                position: 'insideRight'
                            }
                        },
                        itemStyle:{ normal:{color:'#f47209'} },
                        data: json.titleCount
                    }

                ]
            };
            myChart.setOption(option);
        }
        function titleSum(data){
            var option = {
                backgroundColor: '#fbfbfb',              //背景色
                title:{
                    text: '新闻话题曝光量【实时】统计',
                    subtext: '数据来自搜狗实验室'
                },
                tooltip :{
                    formatter: "{a} <br/>{b} : {c}%"
                },
                toolbox:{
                    feature:{
                        restore: {},
                        saveAsImage: {}
                    }
                },
                series:[
                    {
                        name: '业务指标',
                        type: 'gauge',
                        max:50000,
                        detail: {formatter:'{value}个话题'},
                        data: [{value: 50,name: '话题曝光量'}]
                    }
                ]
```

```
            };
            option.series[0].data[0].value = data;
            myChart_sum.setOption(option,true);
        }
    </script>
    </body>
</html>
```

14.5 系统运行测试

(1) 启动 HBase,新建 weblogs 表。新建表语句如下:

create 'weblogs','info'

(2) 启动 MySQL,新建 webCount 表。新建表语句如下:

```
create database test;
use test;
CREATE TABLE `webCount`(
    `titleName` varchar(255) DEFAULT NULL,
    `count` int(11)DEFAULT NULL
) ENGINE = InnoDB DEFAULT CHARSET = utf8;
```

(3) 处理搜狗数据集,将文件中的 Tab 和空格替换成逗号,在日志采集端执行以下 CentOS 命令:

```
cat weblog|tr "\t" "," > weblog1.log
cat weblog1.log|tr " " "," > weblog.log
```

(4) 在日志采集端运行模拟日志生成程序 weblog.sh,weblog.sh 运行结果如图 14-5 所示。

图 14-5 weblog.sh 运行结果

(5) 在日志聚合节点上启动 Flume 聚合脚本 flume-start.sh,将采集的数据分发到 Kafka 集群和 HBase,日志聚合节点 flume-start.sh 运行结果如图 14-6 所示。

(6) 在日志采集节点启动 Flume 采集脚本 flume-start.sh,将采集的数据发送到

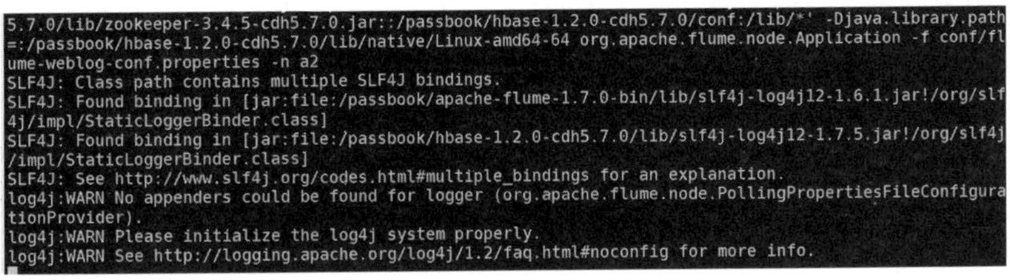

第14章 实战案例——新闻话题实时统计分析系统

图 14-6　日志聚合节点 flume-start.sh 运行结果

Flume 日志聚合节点。

（7）启动 Kafka，并创建业务数据 Topic。语句如下：

```
bin/kafka-server-start.sh config/server.properties &
bin/kafka-topics.sh --create --zookeeper localhost:2181 --topic weblogs --partitions 1 --replication-factor 1
```

（8）启动 StructuredStreamingKafka，实现每 5s 从 Kafka 中读取数据存储到关系数据库 MySQL 中，StructuredStreamingKafka 运行界面如图 14-7 所示。

图 14-7　StructuredStreamingKafka 运行界面

（9）发布 Web 应用，访问大屏显示页面 index.html，大屏显示页面如图 14-8 所示。

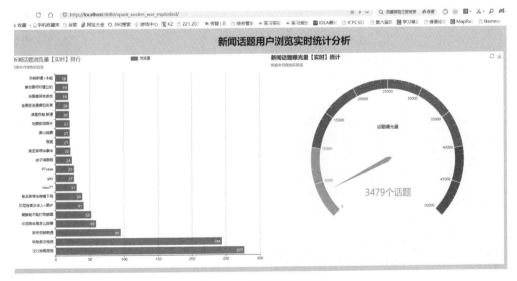

图 14-8　大屏显示页面

14.6 本章小结

本章介绍了新闻话题实时统计分析系统的开发核心思路和核心代码以及启动调用过程，由于系统比较庞大，代码较多，本书只列出了系统的基本构建和核心思路与代码。

参 考 文 献

[1] 徐宗本.用好大数据须有大智慧——准确把握、科学应对大数据带来的机遇和挑战[J].党政干部参考,2016,000(007):8-10.
[2] 林子雨,赖永炫,陶继平.Spark 编程基础(Scala 版)[M].北京:人民邮电出版社,2018.
[3] 舍恩伯格.大数据时代:生活、工作、思维的大变革[M].杭州:浙江人民出版社,2013.
[4] 林子雨.大数据基础编程、实验和案例教程[M].北京:清华大学出版社,2017.
[5] 周志华.机器学习[M].北京:清华大学出版社,2016.
[6] 黑马程序员.Hadoop 大数据技术原理与应用[M].北京:清华大学出版社,2019.
[7] WHITE T.Hadoop 权威指南[M].王海,华东,刘喻,等译.北京:清华大学出版社,2017.
[8] 艾叔.Spark 大数据编程实用教程[M].北京:机械工业出版社,2020.
[9] 中国计算机学会大数据专家委员会.2013 年中国大数据技术与产业发展白皮书[R].北京:中国计算机学会,2013.
[10] 中国计算机学会大数据专家委员会.2019 年中国大数据技术与产业发展白皮书[R].北京:中国计算机学会,2019.
[11] 赵国栋,易欢欢,糜乃军,等.大数据时代的历史机遇:产业变革与数据科学[M].北京:清华大学出版社,2013.
[12] 程学旗,靳小龙,杨婧,等.大数据技术进展与发展趋势[J].科技导报,2016,34(14):49-59.

图书资源支持

感谢您一直以来对清华版图书的支持和爱护。为了配合本书的使用,本书提供配套的资源,有需求的读者请扫描下方的"书圈"微信公众号二维码,在图书专区下载,也可以拨打电话或发送电子邮件咨询。

如果您在使用本书的过程中遇到了什么问题,或者有相关图书出版计划,也请您发邮件告诉我们,以便我们更好地为您服务。

我们的联系方式:

地　　址:北京市海淀区双清路学研大厦 A 座 714

邮　　编:100084

电　　话:010-83470236　010-83470237

客服邮箱:2301891038@qq.com

QQ:2301891038(请写明您的单位和姓名)

资源下载: 关注公众号"书圈"下载配套资源。

资源下载、样书申请

书圈

获取最新书目

观看课程直播